Silicon-Based Nanomaterials

Silicon-Based Nanomaterials: Technology and Applications

Special Issue Editor

Robert W. Kelsall

MDPI • Basel • Beijing • Wuhan • Barcelona • Belgrade

MDPI

Special Issue Editor
Robert W. Kelsall
University of Leeds
UK

Editorial Office
MDPI
St. Alban-Anlage 66
4052 Basel, Switzerland

This is a reprint of articles from the Special Issue published online in the open access journal *Nanomaterials* (ISSN 2079-4991) in 2018 (available at: https://www.mdpi.com/journal/nanomaterials/special_issues/Silicon_Based_Nanomaterials)

For citation purposes, cite each article independently as indicated on the article page online and as indicated below:

LastName, A.A.; LastName, B.B.; LastName, C.C. Article Title. *Journal Name* **Year**, *Article Number*, Page Range.

ISBN 978-3-03921-042-8 (Pbk)
ISBN 978-3-03921-043-5 (PDF)

Cover image courtesy of Vasileios Nika and Spyros Gallis.

Contents

About the Special Issue Editor

Robert W. Kelsall graduated with BSc (Hons) and Ph.D. degrees from the University of Durham in 1985 and 1989, respectively. His doctoral research involved the development of Monte Carlo simulations of electronic transport in GaAs quantum wells. He joined the University of Leeds in 1993, where he is currently Professor of Semiconductor Nanotechnology and Head of the School of Electronic & Electrical Engineering. His research interests are in the simulation and design of nanostructured electronic and photonic semiconductor devices. He has published over 350 papers in international journals and at both national and international conferences. Professor Kelsall is a Fellow of the UK Institute of Physics.

Preface to "Silicon-Based Nanomaterials: Technology and Applications"

Silicon has been proven to be remarkably resilient as a commercial electronic material. The microelectronics industry has harnessed nanotechnology to continually push the performance limits of silicon devices and integrated circuits. Rather than shrinking its market share, silicon is displacing "competitor" semiconductors in domains such as high-frequency electronics and integrated photonics. There are strong business drivers underlying these trends; however, an important contribution is also being made by research groups worldwide, who are developing new configurations, designs, and applications of silicon-based nanoscale and nanostructured materials. This Special Issue features a selection of papers which illustrate recent advances in the preparation of chemically or physically engineered silicon-based nanostructures and their application in electronic, photonic, and mechanical systems.

Robert W. Kelsall
Special Issue Editor

nanomaterials

MDPI

Review
Optical Properties of Tensilely Strained Ge Nanomembranes

Roberto Paiella [1,*] and Max G. Lagally [2]

[1] Department of Electrical and Computer Engineering and Photonics Center, Boston University, Boston, MA 02215, USA
[2] Department of Materials Science and Engineering, University of Wisconsin, Madison, WI 53706, USA; lagally@engr.wisc.edu
* Correspondence: rpaiella@bu.edu; Tel.: +1-617-353-8883

Received: 16 April 2018; Accepted: 4 June 2018; Published: 6 June 2018

Abstract: Group-IV semiconductors, which provide the leading materials platform of microelectronics, are generally unsuitable for light emitting device applications because of their indirect-bandgap nature. This property currently limits the large-scale integration of electronic and photonic functionalities on Si chips. The introduction of tensile strain in Ge, which has the effect of lowering the direct conduction-band minimum relative to the indirect valleys, is a promising approach to address this challenge. Here we review recent work focused on the basic science and technology of mechanically stressed Ge nanomembranes, i.e., single-crystal sheets with thicknesses of a few tens of nanometers, which can sustain particularly large strain levels before the onset of plastic deformation. These nanomaterials have been employed to demonstrate large strain-enhanced photoluminescence, population inversion under optical pumping, and the formation of direct-bandgap Ge. Furthermore, Si-based photonic-crystal cavities have been developed that can be combined with these Ge nanomembranes without limiting their mechanical flexibility. These results highlight the potential of strained Ge as a CMOS-compatible laser material, and more in general the promise of nanomembrane strain engineering for novel device technologies.

Keywords: nanomembranes; optical gain media; group-IV semiconductors; strain engineering

1. Introduction

The development of group-IV semiconductor lasers that can be integrated within complex microelectronic systems with a CMOS-compatible process is a major goal of current research in optoelectronics. The key underlying challenge is the indirect-bandgap nature of the leading materials of Si-based microelectronics (i.e., Si, Ge, and their alloy SiGe), which results in extremely low light-emission efficiencies. The use of Ge as the laser active material is particularly compelling because of the relatively small difference (~0.14 eV) between its direct and fundamental-indirect bandgap energies. Ge is also more generally the subject of extensive renewed research interest [1]. Its large mobility of both electrons and holes is especially attractive in CMOS microelectronics, although challenges still exist related to the choice of gate dielectric and n doping [2]. In optoelectronics, near-infrared Ge photodetectors are already well-established [3].

A promising approach that is currently being pursued to overcome the indirect-bandgap nature of Ge for application to light emitting devices is based on strain engineering. Extensive theoretical work has shown that the introduction of tensile strain (larger than about 4% uniaxial or 1.9% biaxial for the standard (001) crystal orientation) has the effect of converting Ge into a direct-bandgap semiconductor capable of providing optical gain [4–11]. At the same time, the direct-bandgap energy of Ge under tensile strain is also red-shifted into the short-wave mid-infrared spectral region beyond 2 μm, where a wide range of biochemical sensing applications exists that could benefit strongly from an integrated lab-on-a-chip device platform. In passing, we note that the incorporation of Sn in Ge produces an effect similar to that of biaxial tensile strain (i.e., a decrease in the direct conduction-band minimum relative to the indirect valleys, and an overall red-shift in bandgap energy). While the growth of high-quality GeSn films remains quite challenging due to phase separation issues, recent progress has led to the demonstration of optically pumped lasing at cryogenic temperatures [12].

In semiconductor micro- and optoelectronics, strain is traditionally introduced by hetero-epitaxy, i.e., through the growth of sufficiently thin layers of the desired materials on a suitably lattice-mismatched template. In the case of Ge on Si, however, this approach is impractical given the exceedingly large mismatch of about 4.2% between the Si and Ge equilibrium lattice constants, which essentially precludes pseudomorphic growth (and would in any case produce compressively rather than tensilely strained Ge). While tensilely strained Ge can be grown on III-V semiconductor templates such as InGaAs [13,14], the main driver for the development of strained-Ge laser materials is their direct compatibility with Si substrates. As a result, novel straining techniques have been widely investigated in the past few years [15–30], involving mechanically stressed Ge nanostructures mostly derived from Ge-on-insulator (GOI) wafers—i.e., Si substrates coated with a buried oxide layer (BOX) underneath a thin Ge film.

In this context, the use of Ge nanomembranes (NMs) has proved to be particularly appealing as a means to studying the basic structural and optical properties of Ge under large biaxial tensile strain. Semiconductor NMs in general are single-crystal films with lateral dimensions in the μm- or mm-scale and thicknesses as small as several tens of nm, which are typically released from their native substrate via the selective etch of an underlying sacrificial layer. By virtue of their crystalline nature and unusual aspect ratios, these films can combine the exceptional electronic and optical properties of inorganic semiconductors with the extreme mechanical flexibility of soft polymeric materials. As a result, they are promising for a wide range of applications in flexible electronics and optoelectronics, as well as for fundamental studies in strain engineering [31].

This article reviews recent work on mechanically stressed Ge NMs aimed at investigating their basic optical properties and demonstrating the suitability of tensilely strained Ge as a Si-compatible laser gain medium [17,20,22,23,28]. The electronic band structure and calculated optical gain spectra of strained Ge are described in Section 2. In Section 3, we briefly review various straining techniques that have been applied to Ge nanostructures, with emphasis on the use of NMs to obtain wide-area biaxial tensile strain. Section 4 is focused on our experimental work based on the latter approach, including the demonstration of strain-enhanced photoluminescence, population inversion under optical pumping, and the formation of direct-bandgap Ge. Finally, in Section 5, we describe recent efforts aimed at the development of mechanically flexible optical cavities that can be integrated with NM active layers without limiting their mechanical flexibility.

2. Electronic and Optical Properties of Strained Ge

Unstrained Ge is an indirect-bandgap semiconductor, where the absolute minimum of the conduction band and the maxima of the valence bands occur, respectively, at the L and Γ points of reciprocal space (Figure 1a). At the same time, the conduction band also features a local minimum at Γ. The introduction of tensile strain in this material has the effect of lowering all conduction-band minima, but at different rates, with the Γ minimum decreasing more rapidly with increasing strain compared to the L valleys. When the strain exceeds about 1.9% (biaxial), the Γ and L minima cross over and Ge becomes a direct-bandgap semiconductor (Figure 1b). At the same time, the degeneracy of the valence bands at the Γ point is also lifted, with the light-hole (LH) band pushed up in energy relative to the heavy-hole (HH) one. The resulting variations in bandgap energies between all the conduction- and valence-band edges just discussed, as computed by deformation-potential theory [32], are plotted in Figure 1c.

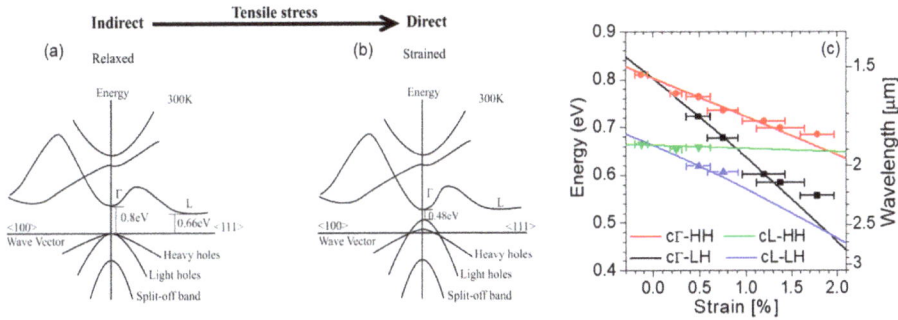

Figure 1. Strain-induced modifications of the Ge band structure. (**a**) Schematic band structure of unstrained Ge; (**b**) Schematic band structure of Ge under 1.9% biaxial tensile strain Reproduced with permission from [23]. American Chemical Society, 2014; (**c**) Solid lines: calculated bandgap energies between the Γ or L conduction-band minima and the HH or LH valence-band maxima of Ge as a function of biaxial strain. Symbols: peak emission energies extrapolated from the strain-resolved luminescence spectra shown in Figure 4a below. Reproduced with permission from [17]. National Academy of Sciences, 2011.

When the direct conduction-band minimum lies below (or at least sufficiently close to) the L valleys, a population inversion at the Γ point can be established under practical pumping conditions. Extensive numerical simulations based on standard models of interband optical transitions in semiconductors [32] indicate that the resulting optical-gain values are adequate for laser operation [20]. Figure 2a shows representative gain spectra of strained Ge(001) due to conduction-to-LH (cΓ-LH) and conduction-to-HH (cΓ-HH) transitions, for both TE and TM polarized light (i.e., with electric field parallel and perpendicular to the plane of the biaxial strain, respectively). As shown in these plots, optical gain is only produced by the cΓ-LH transitions, which is a consequence of the aforementioned lifting of the valence-band degeneracy under tensile strain, which pushes the LH band above the HH one. As a result, in the presence of external carrier injection, the hole population in tensily strained Ge mostly resides in the LH band. Due to the well-established polarization selection rules of interband transitions in semiconductors [32], the oscillator strength of cΓ-LH transitions is significantly larger for TM-polarized light, as also observed in Figure 2a.

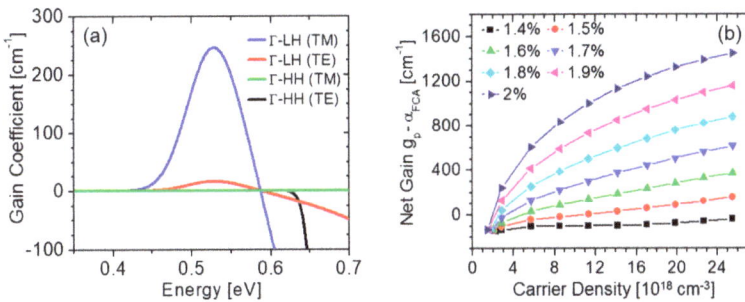

Figure 2. Optical gain properties of tensily strained Ge. (**a**) Calculated TE and TM optical-gain spectra due to cΓ-LH and cΓ-HH transitions in Ge under 1.78-% biaxial tensile strain, in the presence of a density of injected carriers of 3.9×10^{18} cm^{-3}. These values of strain and carrier density correspond to the experimental data shown in Figure 4b below. Reproduced with permission from [20]. John Wiley and Sons, 2012; (**b**) Peak gain coefficient of undoped Ge (including the effect of free-carrier absorption) plotted as a function of injected carrier density for different strain values. Reproduced with permission from [28]. AIP Publishing LLC, 2016.

The peak gain coefficient g_p increases with both carrier density and tensile strain, as illustrated in Figure 2b [28]. To evaluate the full impact of increasing the carrier density N in tensilely strained Ge, this figure shows the net gain g_p-α_{FCA}, where the free-carrier absorption coefficient α_{FCA} also increases (linearly) with N. For the NM cavity geometry described in Section 5 below, the threshold gain required for lasing is estimated to be a few 100 cm^{-1} (if all relevant loss mechanisms are properly minimized). The simulation results of Figure 2b therefore indicate that such a threshold can be reached for strain values larger than about 1.6% biaxial. The larger the strain, the smaller the required density of injected carriers, as expected based on the corresponding decrease in direct versus indirect bandgap energy.

3. Ge-Nanostructure Straining Techniques

The Ge NMs used in the work reviewed below are fabricated by releasing the Ge template layer of a commercial GOI wafer [33]. Specifically, the membrane boundaries and etchant access holes are first patterned in this Ge layer using photolithography and reactive ion etching. Next, the underlying BOX layer is completely removed with a wet etch in a hydrofluoric-acid solution. The released Ge NMs are then bonded onto flexible films of polyimide (PI) using a wet transfer procedure. For the strain-dependent measurements described below, the PI films with the attached NMs are integrated in a metallic chamber which is then pressurized with a controlled gas inflow. In this configuration, the NM effectively sits on the surface of an expanding sphere, so that the resulting strain is fully biaxial. A schematic illustration of the experimental setup (with an optical micrograph of a NM bonded onto a PI film) and a photograph of the pressure cell are shown in Figure 3a,b, respectively.

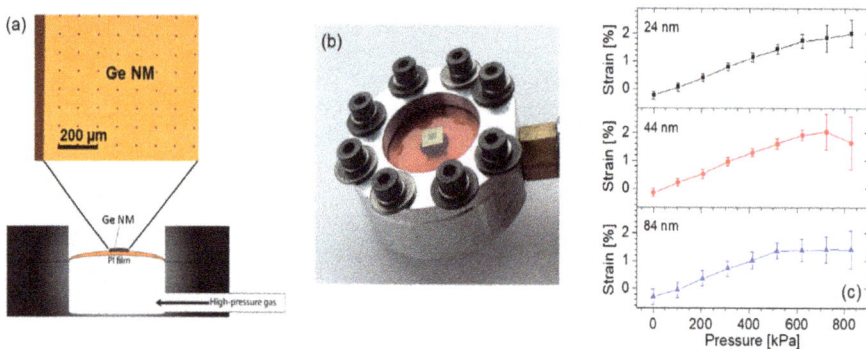

Figure 3. Mechanically stressed Ge NMs. (**a**) Schematic illustration of the experimental setup used to introduce biaxial tensile strain in Ge NMs. A top-view optical micrograph of a NM bonded onto a PI film is also shown. The periodic black dots are etchant access holes, used to facilitate the Ge NM release from its GOI substrate. Reproduced with permission from [23]. American Chemical Society, 2014; (**b**) Photograph of the sample mount under an applied gas pressure of 520 kPa. The Ge NM (here containing an overlaying pillar array, as described in Section 2 below) is visible near the center of the PI film. Reproduced with permission from [28]. AIP Publishing LLC, 2016; (**c**) Strain/stress curves measured by Raman spectroscopy on Ge NMs with three different thicknesses. Reproduced with permission from [17]. National Academy of Sciences, 2011.

The key property of NMs that allows them to reach particularly large strain levels before the onset of plastic deformation is provided by their nanoscale thicknesses. This idea is confirmed and quantified by the Raman measurement results of Figure 3c [17], where the strain introduced in the Ge lattice with the setup of Figure 3a,b is measured as a function of applied gas pressure for three different NM thicknesses. In all three samples, the strain initially increases linearly with applied stress, and then begins to saturate because of the formation of cracks that allow for local strain relaxation in their immediate vicinity. Importantly, the thinner the NM, the larger the strain threshold for the formation of such cracks. In particular, for the smallest NM thickness considered in these measurements (24 nm), a nearly linear strain/stress relation is observed up to a maximum strain value of over 2% (averaged over several random sites on the NM), where Ge is expected to feature a direct fundamental bandgap.

The straining procedure just described is particularly convenient for the purpose of material characterization studies, as it produces large, uniform, and dynamically tunable strain across macroscopic sample areas. Therefore, it provides an ideal platform to study the optical properties of strained Ge reviewed in this article. For practical applications, similar results can in principle be obtained with integrated NM devices on Si chips using MEMS technology [8], piezoelectrics [34], or microfluidics [35,36].

A wide range of other straining techniques has also been investigated. In some of the earliest work in this area, tensile strain has been introduced in plastically relaxed Ge films grown on Si by taking advantage of the different thermal expansion coefficients of Si and Ge with an annealing process [37]. While the resulting strain is quite small (less than about 0.3%), this approach has been used in conjunction with degenerate *n*-type doping (to populate the Γ valley) to demonstrate electrically pumped Ge lasers [38]. To date, however, the performance of these devices remains quite limited, likely due to the excessive free-carrier absorption and Auger recombination losses caused by the degenerate doping.

More recently, significant research efforts have also focused on the use of stressor layers with different types of Ge nanostructures [18,19,21,24–27,29,30]. Typically, these layers consist of material deposited under large compressive strain, which is then allowed to relax partially via elastic strain sharing with an attached partially suspended Ge film. As a result, tensile strain is introduced in the Ge film. Specific stressor materials that have been employed for this purpose include Si_3N_4 [18,25,27,29] and tungsten [19], deposited on a suitably patterned Ge film. Another approach is based on the small (~0.2–0.3%) tensile strain that exists in Ge films grown on Si (after the aforementioned annealing process) or in the Ge template layers of typical GOI substrates. If constricted structures (e.g., in the shape of suspended microbridges) are then patterned in this Ge layer, their local strain can be significantly amplified compared to the surrounding regions, because stress is inversely proportional to cross-sectional area. With this approach, uniaxial and biaxial strain values up to 3.1% [21] and 1.9% [26], respectively, have been measured in micron-scale regions.

These techniques based on built-in stress are directly compatible with on-chip integration, and device applications are already being explored. In [27], high Q-factor Ge microdisk cavities (combining a Si_3N_4 stressor layer with circular Bragg reflectors) have been developed, leading to the observation of line narrowing under optical pumping. A similar microdisk geometry (but without the Bragg reflectors) has also been reported with GeSn [29] as a way to combine the beneficial effects of both tensile strain and Sn incorporation. Finally, an optically pumped laser operating at cryogenic temperatures has also been demonstrated recently [30], based on the suspended-microbridge geometry just described combined with a pair of distributed Bragg mirrors.

4. Strained-Enhanced Light Emission from Mechanically Stressed Ge Nanomembranes

The light-emission properties of mechanically stressed Ge(001) NMs are illustrated by the photoluminescence (PL) data shown in Figure 4 [17,20]. As the applied biaxial tensile strain is increased, the NM emission spectrum is strongly enhanced and simultaneously red-shifted (Figure 4a). This behavior is of course consistent with the corresponding decrease in the Ge direct-bandgap energy relative to the indirect one, which causes a larger and larger fraction of the injected electrons to relax near the Γ minimum of the conduction band. The measured luminescence spectra can be fitted with multiple Gaussian peaks, corresponding to electronic transitions from the direct and indirect minima of the conduction band to the maxima of the strain-split LH and HH valence bands. The emission photon energies extrapolated with this procedure are in good agreement with the calculated bandgap energies of the same transitions, as illustrated in Figure 1c, where the symbols were obtained from the spectra of Figure 4a.

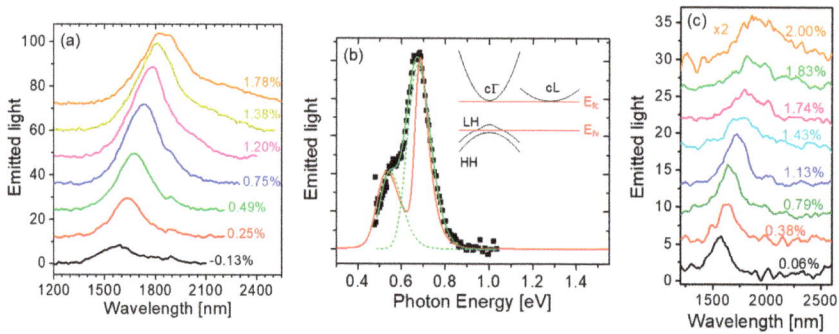

Figure 4. Strain-enhanced light emission from Ge NMs. (**a**) Room-temperature PL spectra of a 40-nm-thick undoped Ge NM at different levels of biaxial tensile strain; (**b**) Symbols: PL spectrum of the NM of (a) (normalized to the spectral response of the measurement setup) at a strain of 1.78%. Green lines: Gaussian fits showing the cΓ-HH and cΓ-LH contributions (dashed) and their sum (solid). Red line: calculated spontaneous emission spectrum. Inset: schematic band diagram of Ge at a strain of 1.78%, and estimated positions of the quasi-Fermi levels relative to the band edges in the presence of the PL pump pulses; (**c**) Room-temperature PL spectra of a 24-nm-thick undoped Ge NM at different levels of biaxial tensile strain. The vertical axes in (**a,c**) are in arbitrary units, and the different spectra are shifted vertically relative to one another for the sake of illustration clarity. The PL pump light in these measurements was provided by an optical parametric oscillator with 5-ns pulse width, 20-Hz repetition rate, 960-nm wavelength, and 3-mW average power. Reproduced with permission from [17]. National Academy of Sciences, 2011.

The same data analysis also indicates that the measured emission spectra of mechanically stressed Ge NMs are dominated by transitions involving HHs, even though under tensile strain the LH band resides at higher energy, and therefore has a higher hole population, compared to the HH band (Figure 1b). This behavior is a result of the aforementioned polarization selection rules: electronic transitions into the LH valence band mostly generate TM-polarized photons that propagate on the plane of the NM and thus cannot be detected in standard surface-emission PL measurements. For the same reason, the full increase in radiative efficiency produced by the applied strain cannot be fully quantified based on these measurement results.

The luminescence spectra measured at high strain also provide evidence of population inversion in the Ge NMs. The key observation in this respect is the ability to resolve in the high-strain emission spectra two separate peaks due to transitions into both LH and HH valence bands near the Γ point (Figure 4b). The relative height of these two peaks depends uniquely on the density of holes injected into the NM under optical pumping. This parameter can therefore be extrapolated by fitting the emission spectra with a standard model of light emission in (strained) semiconductors [32], and then compared to the transparency carrier density of the same NM evaluated with the same model. With this analysis, we find that when the applied strain is above ~1.4% [20], a population inversion is established in the NMs under study.

Finally, Figure 4c shows the strain-resolved PL spectra of the thinnest (24 nm) NM measured in this work. The luminescence intensity of this sample is relatively weak, because of the reduced pump-light absorption and increased nonradiative surface recombination in ultrathin NMs. At the same time, however, the applied strain can be increased to over 2% (where Ge is expected to become a direct-bandgap semiconductor), before the onset of extended-defect formation indicated by the saturation in the strain/stress curve. The emission spectrum measured with this sample at 2% biaxial strain is consistent with the calculated cΓ-HH transition energy in the direct-bandgap regime. Therefore, the results of Figure 4c in conjunction with the Raman data of Figure 3c indicate the formation of direct-bandgap Ge in this NM.

5. Flexible Nanomembrane Optical Cavities

The theoretical and experimental results reviewed in the previous sections substantiate the suitability of mechanically stressed Ge NMs for laser applications. An important challenge in the development of these devices is related to the nanoscale thickness of the NMs, which is too small to provide by itself the required in-plane waveguiding of the (TM-polarized) emitted light. As a result, additional guiding layers must be added to the NMs, without at the same time compromising the NM mechanical flexibility (which is essential to enable gain through straining). This challenge has been addressed with the device geometry shown schematically in Figure 5a, where a periodic array of dielectric columns is deposited on the NM on PI [22,28]. These arrays can be made thick enough and with sufficiently large average refractive index to act as the core guiding layers. At the same time, if their periodicity matches the emission wavelength, they can also provide vertical outcoupling and in-plane optical feedback of the emitted light, by first and second-order diffraction, respectively. Finally, by virtue of their disconnected geometry, the same arrays will not limit the maximum strain that can be introduced in the underlying NM (at least in the regions between neighboring pillars).

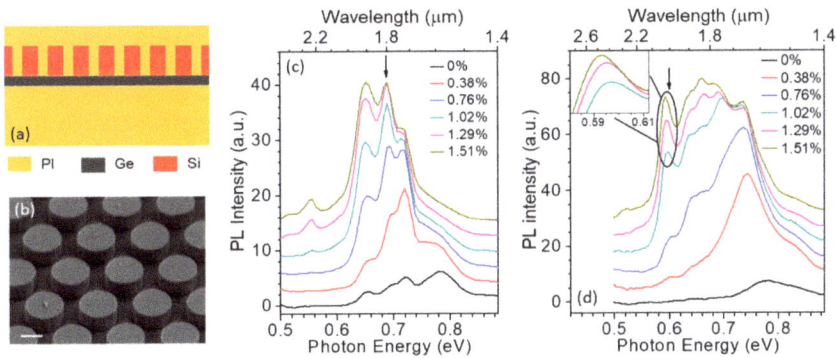

Figure 5. Flexible photonic-crystal cavities for tensilely strained Ge NMs. (**a**) Schematic device cross section; (**b**) SEM image of a Si-pillar array before transfer onto a Ge NM on PI. The scale bar is 500 nm; (**c**,**d**) Strain-resolved room-temperature PL spectra measured on two Ge(001) samples with different column periods *a* and diameters D: (**c**) *a* = 1060 nm, D = 850 nm; (**d**) *a* = 1340 nm, D = 950 nm. Inset of (**d**): zoom-in of the features within the black ellipse, showing strain tuning of the cavity modes. The arrow in each plot indicates the calculated photon energy of the main TM-polarized cavity resonance. (**b**–**d**): Reproduced with permission from [28]. AIP Publishing LLC, 2016.

In recent work, the device geometry just described has been implemented with a novel fabrication process based on direct NM assembly [28]. In this process, an array of Si columns embedded in a PI film is patterned in the template layer of a commercial Si-on-insulator (SOI) wafer, released from its native substrate with a selective wet etch of the SOI BOX layer, and then transferred onto a previously prepared Ge NM on PI. The resulting pillars have extremely smooth sidewalls (Figure 5b) and therefore can be expected to provide minimal optical-scattering losses. Furthermore, because they are based on single-crystal Si, they also feature negligible absorption losses at the strained-Ge emission wavelength.

A large increase and red shift in PL emission with increasing strain is once again observed with these devices, indicating a similar degree of mechanical flexibility as in the bare Ge NMs described above. Representative results are shown in Figure 5c,d for two 50-nm-thick Ge NMs coated with arrays of different periods. In fact, a significantly larger enhancement in PL efficiency is obtained with these devices (up to 12× in Figure 5d) compared to bare NMs, due to the efficient vertical outcoupling of the TM-polarized emission produced by the column array. Furthermore, a complex pattern of relatively narrow features is observed in the PL spectra of Figure 5, associated with different modes

Nanomaterials **2018**, *8*, 407

of the photonic-crystal cavity provided by the pillars. The spectral positions of these features are in good agreement with numerical simulations of the photonic-crystal band structures [28], and can be tuned by varying the array period and the applied strain (inset of Figure 5d). Altogether, these results therefore demonstrate that ultrathin NM active layers can be effectively coupled to an optical cavity while still fully preserving their mechanical flexibility.

6. Conclusions

This article has reviewed recent work, spanning approximately eight years, on mechanically stressed Ge NMs, aimed at the investigation of their ability to sustain large amounts of tensile strain and correspondingly provide enhanced light emission efficiency and optical gain for laser applications. Because of their extreme aspect ratios, these films can be strained beyond the threshold for the formation of direct-bandgap Ge. Numerical simulations indicate that the resulting optical-gain properties are comparable to those of traditional III-V semiconductor laser materials. The expected increase in luminescence efficiency with increasing strain has been measured with NMs of different thicknesses, together with a large red shift of the emission spectra towards the short-wave mid-infrared region. The same experimental results also reveal the presence of population inversion under optical pumping in NMs strained above ~1.4%. Finally, photonic-crystal cavities fully compatible with the flexibility requirements of these NM active layers have been developed, based on a novel membrane-assembly process.

These results in general highlight the potential of strained Ge as a CMOS-compatible laser material. Furthermore, they are promising for the future development of mechanically stressed Ge NM lasers, which could then be integrated on Si chips using MEMS or related technologies. The key remaining challenge is likely related to minimizing the optical losses in the cavity devices of Figure 5, which have not yet shown evidence of spectral narrowing of the emission peaks with increasing pumping and/or strain. In particular, an important loss mechanism appears to be scattering by defects in the NM originating from the GOI template layer, whose structural quality is therefore of critical importance for future progress in this research. In fact, such defects also tend to act as crack initiation sites under mechanical stress, so that their minimization would also allow increasing the maximum achievable strain.

Author Contributions: The research work reviewed in this article was carried out through an extensive collaboration between the two authors (R.P. and M.G.L.). Both authors contributed to the writing of the manuscript.

Acknowledgments: The Ge nanomembrane fabrication and characterization efforts were supported initially by DOE under Grant DE-FG02-03ER46028, and subsequently by AFOSR under Grant FA9550-14-1-0361. The development of the photonic-crystal cavities was supported by NSF under Grant ECCS-1308534. The initial photoluminescence studies were funded by NSF under Grant DMR-0907296. The contribution from several students and research scientists involved in this research at Boston University and the University of Wisconsin—Madison (including Cicek Boztug, Francesca Cavallo, Feng Chen, Xiaorui Cui, RB Jacobson, Debbie Paskiewicz, Jose Sánchez-Pérez, Pornsatit Sookchoo, Faisal Sudradjat, Xiaowei Wang, and Jian Yin) is also gratefully acknowledged.

Conflicts of Interest: The authors declare no conflict of interest.

References

1. Claeys, C.; Simoen, E. *Germanium-Based Technologies: From Materials to Devices*, 1st ed.; Elsevier Science: Oxford, UK, 2007.
2. Kube, R.; Bracht, H.; Chroneos, A.; Posselt, M.; Schmidt, B. Intrinsic and extrinsic diffusion of indium in germanium. *J. Appl. Phys.* **2009**, *106*, 063534. [CrossRef]
3. Michel, J.; Liu, J.; Kimerling, L.C. High-Performance Ge-on-Si Photodetectors. *Nat. Photonics* **2010**, *4*, 527–534. [CrossRef]
4. Soref, R.A.; Friedman, L. Direct-gap Ge/GeSn/Si and GeSn/Ge/Si heterostructures. *Superlattices Microstruct.* **1993**, *14*, 189–193. [CrossRef]

5. Fischetti, M.V.; Laux, S.E. Band structure, deformation potentials, and carrier mobility in strained Si, Ge, and SiGe alloys. *J. Appl. Phys.* **1996**, *80*, 2234–2252. [CrossRef]

6. Menéndez, J.; Kouvetakis, J. Type-I Ge/Ge$_{1-x-y}$Si$_x$Sn$_y$ strained-layer heterostructures with a direct Ge bandgap. *Appl. Phys. Lett.* **2004**, *85*, 1175–1177. [CrossRef]

7. Liu, J.; Sun, X.; Pan, D.; Wang, X.; Kimerling, L.C.; Koch, T.L.; Michel, J. Tensile-strained, n-type Ge as a gain medium for monolithic laser integration on Si. *Opt. Express* **2007**, *15*, 11272–11277. [CrossRef] [PubMed]

8. Lim, P.H.; Park, S.; Ishikawa, Y.; Wada, K. Enhanced direct bandgap emission in germanium by micromechanical strain engineering. *Opt. Express* **2009**, *17*, 16358–16365. [CrossRef] [PubMed]

9. El Kurdi, M.; Fishman, G.; Sauvage, S.; Boucaud, J. Band structure and optical gain of tensile-strained germanium based on a 30 band k.p formalism. *J. Appl. Phys.* **2010**, *107*, 013710. [CrossRef]

10. Aldaghri, O.; Ikonić, Z.; Kelsall, R.W. Optimum strain configurations for carrier injection in near infrared Ge lasers. *J. Appl. Phys.* **2012**, *111*, 053106. [CrossRef]

11. Tahini, H.; Chroneos, A.; Grimes, R.W.; Schwingenschlogl, U.; Dimoulas, A. Strain-induced changes to the electronic structure of germanium. *J. Phys. Condens. Matter.* **2012**, *24*, 195802. [CrossRef] [PubMed]

12. Wirths, S.; Geiger, R.; von den Driesch, N.; Mussler, G.; Stoica, T.; Mantl, S.; Ikonic, Z.; Luysberg, M.; Chiussi, S.; Hartmann, J.M.; et al. Lasing in direct-bandgap GeSn alloy grown on Si. *Nat. Photonics* **2015**, *9*, 88–92. [CrossRef]

13. Huo, Y.; Lin, H.; Chen, R.; Makarova, M.; Rong, Y.; Li, M.; Kamins, T.I.; Vuckovic, J.; Harris, J.S. Strong enhancement of direct transition photoluminescence with highly tensile-strained Ge grown by molecular beam epitaxy. *Appl. Phys. Lett.* **2011**, *98*, 011111. [CrossRef]

14. Jakomin, R.; de Kersauson, M.; El Kurdi, M.; Largeau, L.; Mauguin, O.; Beaudoin, G.; Sauvage, S.; Ossikovski, R.; Ndong, G.; Chaigneau, M.; et al. High quality tensile-strained n-doped germanium thin films grown on InGaAs buffer layers by metal-organic chemical vapor deposition. *Appl. Phys. Lett.* **2011**, *98*, 091901. [CrossRef]

15. El Kurdi, M.; Bertin, H.; Martincic, E.; de Kersauson, M.; Fishman, G.; Sauvage, S.; Bosseboeuf, A.; Boucaud, P. Control of direct band gap emission of bulk germanium by mechanical tensile strain. *Appl. Phys. Lett.* **2010**, *96*, 041909. [CrossRef]

16. Cheng, T.H.; Peng, K.L.; Ko, C.Y.; Chen, C.Y.; Lan, H.S.; Wu, Y.R.; Liu, C.W.; Tseng, H.H. Strain-enhanced photoluminescence from Ge direct transition. *Appl. Phys. Lett.* **2010**, *96*, 211108. [CrossRef]

17. Sánchez-Pérez, J.R.; Boztug, C.; Chen, F.; Sudradjat, F.F.; Paskiewicz, D.M.; Jacobson, R.B.; Lagally, M.G.; Paiella, R. Direct-bandgap light-emitting germanium in tensilely strained nanomembranes. *Proc. Natl. Acad. Sci. USA* **2011**, *108*, 18893–18898. [CrossRef] [PubMed]

18. de Kersauson, M.; El Kurdi, M.; David, S.; Checoury, X.; Fishman, G.; Sauvage, S.; Jacomin, R.; Beaudoin, G.; Sagnes, I.; Boucaud, P. Optical gain in single tensile-strained germanium photonic wire. *Opt. Express* **2011**, *19*, 17925–17934. [CrossRef] [PubMed]

19. Nam, D.; Sukhdeo, D.; Roy, A.; Balram, K.; Cheng, S.; Huang, K.; Yuan, Z.; Brongersma, M.; Nishi, Y.; Miller, D.; et al. Strained germanium thin film membrane on silicon substrate for optoelectronics. *Opt. Express* **2011**, *19*, 25866–25872. [CrossRef] [PubMed]

20. Boztug, C.; Sánchez-Pérez, J.R.; Sudradjat, F.F.; Jacobson, R.B.; Paskiewicz, D.M.; Lagally, M.G.; Paiella, R. Tensilely strained germanium nanomembranes as infrared optical gain media. *Small* **2013**, *9*, 622–630. [CrossRef] [PubMed]

21. Süess, M.J.; Geiger, R.; Minamisawa, R.A.; Schiefler, G.; Frigerio, J.; Christina, D.; Isella, G.; Spolenak, R.; Faist, J.; Sigg, H. Analysis of enhanced light emission from highly strained germanium microbridges. *Nat. Photonics* **2013**, *7*, 466–472. [CrossRef]

22. Boztug, C.; Sánchez Pérez, J.R.; Yin, J.; Lagally, M.G.; Paiella, R. Grating-coupled mid-infrared light emission from tensilely strained germanium nanomembranes. *Appl. Phys. Lett.* **2013**, *103*, 201114. [CrossRef]

23. Boztug, C.; Sánchez Pérez, J.R.; Cavallo, F.; Lagally, M.G.; Paiella, R. Strained-germanium nanostructures for infrared photonics. *ACS Nano* **2014**, *8*, 3136–3151. [CrossRef] [PubMed]

24. Guilloy, K.; Pauc, N.; Gassenq, A.; Gentile, P.; Tardif, S.; Rieutord, F.; Calvo, V. Tensile strained germanium nanowires measured by photocurrent spectroscopy and X-ray microdiffraction. *Nano Lett.* **2015**, *15*, 2429–2433. [CrossRef] [PubMed]

25. Millar, R.W.; Gallacher, K.; Samarelli, A.; Frigerio, J.; Chrastina, D.; Isella, G.; Dieing, T.; Paul, D.J. Extending the emission wavelength of Ge nanopillars to 2.25 μm using silicon nitride stressors. *Opt. Express* **2015**, *23*, 18193–18202. [CrossRef] [PubMed]

26. Gassenq, A.; Guilloy, K.; Osvaldo Dias, G.; Pauc, N.; Rouchon, D.; Hartmann, J.-M.; Widiez, J.; Tardif, S.; Rieutord, F.; Escalante, J.; et al. 1.9% Bi-axial tensile strain in thick germanium suspended membranes fabricated in optical germanium-on-insulator substrates for laser applications. *Appl. Phys. Lett.* **2015**, *107*, 191904. [CrossRef]

27. El Kurdi, M.; Prost, M.; Ghrib, A.; Elbaz, A.; Sauvage, S.; Checoury, X.; Beaudoin, G.; Sagnes, I.; Picardi, G.; Ossikovski, R.; et al. Tensile-strained germanium microdisks with circular Bragg reflectors. *Appl. Phys. Lett.* **2016**, *108*, 091103. [CrossRef]

28. Yin, J.; Cui, X.; Wang, X.; Sookchoo, P.; Lagally, M.G.; Paiella, R. Flexible nanomembrane photonic-crystal cavities for tensilely strained-germanium light emission. *Appl. Phys. Lett.* **2016**, *108*, 241107. [CrossRef]

29. Millar, R.W.; Dumas, D.C.S.; Gallacher, K.F.; Jahandar, P.; MacGregor, C.; Myronov, M.; Paul, D.J. Mid-infrared light emission >3 μm wavelength from tensile strained GeSn microdisks. *Opt. Express* **2017**, *25*, 25374–25385. [CrossRef] [PubMed]

30. Bao, S.; Kim, D.; Onwukaeme, C.; Gupta, S.; Saraswat, K.; Lee, K.H.; Kim, Y.; Min, D.; Jung, Y.; Qiu, H.; et al. Low-threshold optically pumped lasing in highly strained germanium nanowires. *Nat. Commun.* **2017**, *8*, 1845. [CrossRef] [PubMed]

31. Rogers, J.A.; Lagally, M.G.; Nuzzo, R.G. Synthesis, assembly and applications of semiconductor nanomembranes. *Nature* **2011**, *477*, 45–53. [CrossRef] [PubMed]

32. Chuang, S.L. *Physics of Photonic Devices*, 2nd ed.; Wiley: Hoboken, NJ, USA, 2009.

33. Cavallo, F.; Lagally, M.G. Semiconductors turn soft: inorganic nanomembranes. *Soft Matter* **2010**, *6*, 439–455. [CrossRef]

34. Chen, Y.; Zhang, J.; Zopf, M.; Jung, K.; Zhang, Y.; Keil, R.; Ding, F.; Schmidt, O.G. Wavelength-tunable entangled photons from silicon-integrated III–V quantum dots. *Nat. Comm.* **2016**, *7*, 10387. [CrossRef] [PubMed]

35. Rasmussen, A.; Gaitan, M.; Locascio, L.E.; Zaghloul, M.E. Fabrication techniques to realize CMOS-compatible microfluidic microchannels. *J. Microelectromech. Syst.* **2001**, *10*, 286–297. [CrossRef]

36. Zhao, Y.; Zhou, J.; Dai, W.; Zheng, Y.; Wu, H. A convenient platform of tunable microlens arrays for the study of cellular responses to mechanical strains. *J. Micromech. Microeng.* **2011**, *21*, 054017. [CrossRef]

37. Cannon, D.D.; Liu, J.; Ishikawa, Y.; Wada, K.; Danielson, D.T.; Jongthammanurak, S.; Michel, J.; Kimerling, L.C. Tensile strained epitaxial Ge films on Si(100) substrates with potential application in L-band telecommunications. *Appl. Phys. Lett.* **2004**, *84*, 906–908. [CrossRef]

38. Camacho-Aguilera, R.E.; Cai, Y.; Patel, N.; Bessette, J.T.; Romagnoli, M.; Kimerling, L.C.; Michel, J. An electrically pumped germanium laser. *Opt. Express* **2012**, *20*, 11316–11320. [CrossRef] [PubMed]

nanomaterials

MDPI

Article

Enhanced Electroluminescence from Silicon Quantum Dots Embedded in Silicon Nitride Thin Films Coupled with Gold Nanoparticles in Light Emitting Devices

Ana Luz Muñoz-Rosas [1], Arturo Rodríguez-Gómez [2,*] and Juan Carlos Alonso-Huitrón [3]

[1] Centro de Ciencias Aplicadas y Desarrollo Tecnológico, Universidad Nacional Autónoma de Mexico, A.P. 70-180, Ciudad de Mexico 04510, Mexico; analu.mrosas@gmail.com

[2] Instituto de Física, Universidad Nacional Autónoma de Mexico, Circuito de la Investigación Científica s/n, Ciudad Universitaria, A.P. 20-364, Coyoacán, Ciudad de Mexico 04510, Mexico

[3] Instituto de Investigaciones en Materiales, Universidad Nacional Autónoma de México, Ciudad Universitaria, A.P. 70-360, Coyoacán, Ciudad de Mexico 04510, Mexico; alonso@unam.mx

* Correspondence: arodriguez@fisica.unam.mx

Received: 24 February 2018; Accepted: 19 March 2018; Published: 22 March 2018

Abstract: Nowadays, the use of plasmonic metal layers to improve the photonic emission characteristics of several semiconductor quantum dots is a booming tool. In this work, we report the use of silicon quantum dots (SiQDs) embedded in a silicon nitride thin film coupled with an ultra-thin gold film (AuNPs) to fabricate light emitting devices. We used the remote plasma enhanced chemical vapor deposition technique (RPECVD) in order to grow two types of silicon nitride thin films. One with an almost stoichiometric composition, acting as non-radiative spacer; the other one, with a silicon excess in its chemical composition, which causes the formation of silicon quantum dots imbibed in the silicon nitride thin film. The ultra-thin gold film was deposited by the direct current (DC)-sputtering technique, and an aluminum doped zinc oxide thin film (AZO) which was deposited by means of ultrasonic spray pyrolysis, plays the role of the ohmic metal-like electrode. We found that there is a maximum electroluminescence (EL) enhancement when the appropriate AuNPs-spacer-SiQDs configuration is used. This EL is achieved at a moderate turn-on voltage of 11 V, and the EL enhancement is around four times bigger than the photoluminescence (PL) enhancement of the same AuNPs-spacer-SiQDs configuration. From our experimental results, we surmise that EL enhancement may indeed be due to a plasmonic coupling. This kind of silicon-based LEDs has the potential for technology transfer.

Keywords: silicon quantum dots; localized surface plasmon resonances; light emitting devices; gold nanoparticles; electroluminescence enhancement

1. Introduction

Since the first reports of luminescence and electroluminescence, originated by quantum size effects, from highly confined silicon materials (superlattices, quantum dots, and quantum wires) [1–3], there has been a growing interest in the development of monolithic silicon photonics as the optical analogue of silicon microelectronics [4–7]. In order to meet this goal, arduous work has been done over the years to fabricate light emitters and electroluminescent devices based mainly on crystalline and amorphous silicon quantum dots (SiQDs) embedded in silicon nitride and silicon dioxide films, and to tune the photoluminescence by controlling the size and the surface passivation of the SiQDs [8–13]. However, the silicon photonics have evolved slowly, mainly because the illumination efficiency from confined silicon is still very low compared with that of the direct-band gap III–V semiconductors [7].

With the purpose of increasing the efficiency and controlling the light emission from SiQDs, several research groups have used different configurations of noble metals (Au, Ag) nanoparticles and nanostructures in the vicinity of silicon quantum dots. One of the first pioneer works, reported local field-enhanced light emission from silicon nanocrystals implanted in quartz close to a surface film of nanoporous gold, prepared by wet chemical methods [14]. In other posterior works, authors have reported enhanced luminescence from SiQDs implanted in quartz and coupled to Ag island arrays fabricated by electron beam lithography or by subsequent implantation. Those enhancements are reported at emission frequencies that correspond to the collective dipole plasmon resonances [15–18]. In this regard, we recently reported photoluminescence (PL) enhancement from thin films of SiQDs embedded in silicon nitride coupled to a monolayer of Au nanoparticles which are separated by a nanometric dielectric silicon nitride thin film [19]. In that article, the films were deposited using dry and low-temperature techniques highly compatible with the pre-existing silicon microelectronics technology, such as remote plasma enhanced chemical vapor deposition (RPECVD) and direct current (DC) sputtering.

A shared feature in the coupled structures reported in all these previous works is that all of them have a spacer between the metal and the SiQDs. In turn, the spacer must be a dielectric, and must have a well-defined thickness ranging from 10 to 20 nm; when these two last conditions are not met, then the metal-spacer-SiQDs structure, far from improving its PL, decreases it [17]. All these works also shared the proposal that the PL enhancement was due to effects of localized surface plasmon resonance (LSPR) [16–18,20]. Therefore, they have motivated the development of other plasmonic coupled systems with enhanced PL, such as core-shell-type SiQDs-based nanocomposites consisting of a Au nanoparticle (NP) core and a thick shell of SiQDs agglomerates [21], a structure where SiQDs are placed in a gap between a gold thin film and an Au nanoparticle [22]. Or finally, a structure composed of a monolayer of luminescent SiQDs and a silver (Ag) film over nanosphere (AgFON) plasmonic structure, separated with a polymer spacer [23].

The same approach of coupling SiQDs to localized surface plasmons (LSP) has also been applied to enhance the electroluminescence of electroluminescent devices based on SiQDs embedded in silicon-rich silicon nitride (SiN$_x$(SiQDs)) films deposited by plasma enhanced chemical vapor deposition (PECVD) [24–26]. In these works, the enhancement of the electroluminescence by LSP was investigated in two different types of light emitting devices and/or diodes (LEDs) with layered structures, such as ITO-Ni-Au(transparent electrode)/SiN$_x$(SiQDs)/Ag islands/p/p$^+$-Si(substrate)/Ni-Au(back electrode) [24], ITO/SiO$_2$/SiN$_x$(SiQDs)/Ag islands/p/p$^+$-Si/Al [25,26]. In both types of layered LEDs, the large enhancement of the electroluminescence (EL) (up to 434% relative to SiQD LED without an Ag layer) was attributed to the Ag island layer, which gives rise to an increase in the radiative efficiency as a result of SiQDs-LSP coupling. Additionally, it was also observed, an increase in the current injection efficiency through improved carrier tunneling between the rough surface of the Ag layer and the SiQDs. It is also worth mentioning that in the case of the second type of layered LEDs, the effect of the SiO$_2$ layer was not discussed.

An interesting work that questions the mechanisms of the enhancement of the spontaneous emission rate due to the LSP coupling between the SiQDs excitons and the Ag island layer, is the one authored by Baek Kim and collaborators. There, they report an enhancement of 493% in the EL of a LED with a layered structure NiO-Ni-Au(transparent electrode)/SiN$_x$(SiQDs)/rough p$^+$-Si, which do not contain any Ag layer, but instead, it was fabricated on a nano-roughened Si substrate [27]. In a recent work, novel electroluminescent structures were fabricated using SiQDs/SiO$_2$ multilayers fabricated by PECVD on a Pt nanoparticle—sputtering-coated Si nanopillar array substrate, with ITO as the transparent electrode. The electroluminescence enhancement observed in these EL structures was attributed to both, the possible resonance coupling between the localized surface plasmon (LSP) of Pt NPs, and the band-gap emission of SiQDs/SiO$_2$ multilayers, and the surface roughening originated by the nanopillar array [28].

In this work, we have investigated the electroluminescence of four different configurations of metal insulator semiconductor (MIS)-type nano-layered structures using SiQDs embedded in silicon nitride luminescent films and dielectric silicon nitride (as spacer), both deposited by remote PECVD (RPECVD), and gold nanoparticles (AuNPs) deposited by sputtering on p-type silicon substrate. We found that there is a maximum EL enhancement when the appropriate AuNPs-spacer-SiQDs configuration is used. Furthermore, it was identified that the EL enhancement is around four times bigger than the PL enhancement previously observed in an identical nano-layered configuration [19]. From this work, we can conclude that the EL enhancement may indeed be due to a plasmonic coupling. Nevertheless, we also identify that the presence of gold nanoparticles in the EL device allow a more efficient distribution of charge carriers towards the luminescent centers (SiQDs). Consequently, we confirm that more than one mechanism could be involved in the optimized electroluminescence.

2. Methods

To investigate the effect of gold nanoparticles in the vicinity of silicon quantum dots on the electroluminescence of the fabricated light emitting devices (LED), we used p-type silicon wafers (100), with a concentration of 10×10^{15} holes, as the semiconductor substrate. Common solvent cleaning was employed for all samples, and additionally a standard cleaning 1 as the widely reported by Radio Corporation of America (RCA clean) was used for silicon wafers. Gold nanoparticles (AuNPs) were deposited using a Cressington 108 Sputter Coater (TED PELLA, INC., Redding, CA, USA) in 0.8 mb argon atmosphere. Quartz substrates were selectively used to make depositions of the different films in order to measure their absorption spectrum.

Silicon nitride films with different thicknesses and compositions were deposited using a remote plasma enhanced chemical vapor deposition (RPECVD) system whose characteristics have been reported elsewhere [29]. A substrate temperature of 300 °C, radio frequency power of 150 watts and pressure of the reaction chamber of 300 mT were used as deposition parameters. The flow rates of H_2, Ar, and SiH_2Cl_2 were 10, 75, and 5 sccm, respectively for all the deposited films. A NH_3 flow rate of 600 sccm was settled to attain a non-radiative silicon nitride (SiN_x) insulating film with NH_3/SiH_2Cl_2 gas flow ratio of $R = 120$. In addition, a NH_3 flow rate of 200 sccm was used to obtain the radiative silicon rich silicon nitride film (SiQDs) with a NH_3/SiH_2Cl_2 gas flow ratio of $R = 40$.

Four different types of LED structures were fabricated with the aim of obtaining separate information on the role of the AuNPs and silicon nitride layers in the enhancement of the electroluminescence of the devices. Two EL structures were fabricated without AuNPs, and these were considered as reference structures. The first reference EL structure was fabricated by depositing an 80 ± 5 nm thick silicon-rich silicon nitride (SiQDs) on the surface of the p-type silicon substrate, and then a ZnO-Al film, resulting in the (p-Si/SiQDs (80 ± 5 nm)/ZnO-Al) structure, named PR1 (see Figure 1a). The second reference structure was fabricated as the previous PR1 structure, but depositing before the SiQDs film a 10 ± 2 nm thick non-radiative silicon nitride (SiN_x) film. This (p-Si/SiN_x (10 ± 2 nm)/SiQDs (80 ± 5 nm)/ZnO-Al) reference structure was named PR2 (see Figure 1c). On the other hand, the first AuNPs-enhanced electroluminescent structure was fabricated also as the PR1 structure but depositing a layer of AuNPs before the SiQDs film as shown in Figure 1b. This (p-Si/AuNPs/ SiQDs (80 ± 5 nm)/ZnO-Al) structure was named P1. The second AuNP-enhanced electroluminescent structure was fabricated as the PR2 structure, but depositing a layer of AuNPs before the non-radiative silicon nitride (SiN_x) film. This (p-Si/AuNPs/SiN_x (10 ± 2 nm)/SiQDs (80 ± 5 nm)/ZnO-Al) was named P2 (see Figure 1d). The different silicon nitride layers were grown into the same chamber without exposition of the films to the ambient atmosphere by changing the NH_3 gas flow rate. As transparent conductive contact (TCC), aluminum doped zinc oxide (ZnO-Al) was deposited by ultrasonic spray pyrolysis on top of the silicon-rich silicon nitride layer, defining square patterns with sides of 2 mm. Finally, an aluminum metal layer of 100 nm was used as the bottom electrode, deposited by vacuum evaporation for all the devices.

The presence of the SiQDs and Au nanoparticles were characterized by high-resolution transmission electron microscopy (HRTEM) using a field emission gun JEM-2010F microscope (JEOL, INC., Peabody, MA, USA) operating at 200 kV. The HRTEM images were digitally treated with a GATAN micrograph system (GATAN, INC., Pleasanton, CA, USA). The thickness (Th) of the films was measured using a manual Gaertner 117 ellipsometer (Gaertner Scientific Coroporation, Chicago, IL, USA) equipped with a He-Ne laser (632 nm). Ultraviolet-visible (UV-vis) transmission measurements were carried out in the range from 300 to 1100 nm using a double-beam PerkinElmer Lambda 35 UV-vis spectrophotometer (PerkinElmer, Billerica, MA, USA). The chemical composition of the layers was determined by Fourier transform infrared spectroscopy (FTIR) by using a Nicolet 6700 system (Nicolet Instrument Corporation, Madison, WI, USA). X-ray photoelectron spectroscopy (XPS) depth profiles were performed in an ultra-high vacuum system scanning XPS microprobe PHI 5000 VersaProbe II (PHYSICAL ELECTRONICS, Inc., Chanhassen, MN, USA). Photoluminescence (PL) and electroluminescent (EL) measurements were carried out in a dark room at room temperature. PL spectra were obtained using an unfocused beam of 25 mW from a Kimmon He-Cd laser operating at 325 nm (3.81 eV) (Kimmon Electric US, Ltd., Englewood, CO, USA). An 2200-72-1 power supply (Keithley Instrumemts, Cleveland, OH, USA) source and Digital Multimeter Tektronix DMM4050 (Tektronix, Beaverton, OR, USA) were used as power source and current meter, respectively. The PL and EL spectra were recorded with a Fluoromax-Spex spectrofluorometer (SPEX Industries, Edison, NJ, USA) at room temperature. Finally, a field emission-scanning electron microscope (JEOL7600F FE-SEM) (JEOL, Inc., Peabody, MA, USA) was used to observe the cross-section of the Metal-Insulator-Semiconductor (MIS-type devices).

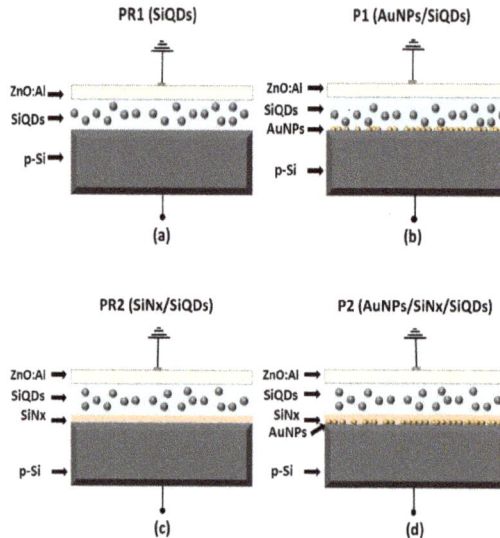

Figure 1. (a–d) Schematic representation of the fabricated structures PR1, P1, PR2, and P2.

3. Results and Discussion

3.1. Preparation of the Layered Luminescent Devices

To characterize the chemical and physical properties of the SiQDs and SiN$_x$ films used for the fabrication of the devices, similar thicknesses of each these layers were deposited on silicon and quartz (Table 1). From Tauc plots [30], the band gap of the SiQDs and SiN$_x$ films were obtained and are presented in Table 1, which are 4.04 and 4.68 eV, respectively. As expected, the SiN$_x$ layer with $R = 120$ has a wider band gap than the SiQDs layer with $R = 40$, and closer to that of the stoichiometric Si$_3$N$_4$

(5 eV) [31,32]. The latter is in agreement with those works which have demonstrated that a higher NH_3 flow rate, as one of the precursors for the deposition of silicon nitride films, increases its band-gap energy, since Si atoms are bonded to more N atoms, due to the larger electronegativity of N compared to that of Si and H [33]. Then, the gap energy of the SiQDs layer should be lower due to its higher content of silicon, whose band gap is 1.1 eV [31,32]. Furthermore, the refractive index of the SiQDs film obtained by null ellipsometry was slightly higher than that of the SiN_x layer, owing to its higher silicon content. However, it is worth noting the low value of the refractive index of the SiN_x film compared to the stoichiometric Si_3N_4 value of 2. A similar trend of the refractive index has been observed in silicon nitride films when increasing the flow rate of the NH_3 precursor, probably due to an increase in the H and N content into the films [34,35].

The Fourier transform infrared spectra of the SiQDs and SiN_x films are depicted in Figure 2. In both samples, the characteristic Si–N (840 cm^{-1}), N–H (1180 cm^{-1}), and N–H (3350 cm^{-1}) bands of silicon nitride are found. However, the Si–H (2190 cm^{-1}) band is only clearly observed in the SiQDs layer. The lack of this band in the SiN_x film with a higher NH_3 flow rate should be due to the incorporation of nitrogen atoms in silicon sites of the Si–H groups, which is consistent with those works using a wide range of this precursor gas [33,35].

Table 1. Some physical properties of the silicon nitride layers with different chemical composition.

$R = NH_3/SiH_2Cl_2$ Gas Flow Ratio	Sample	Thickness (nm)	Refractive Index	Optical Band Gap (eV)
120	SiN_x	96.5	1.78	4.68
40	SiQDs	97.9	1.84	4.04

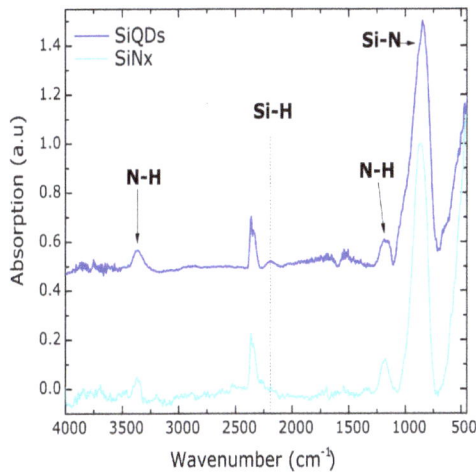

Figure 2. FTIR spectra for the two different NH_3 flow rates used to attain silicon quantum dots (SiQDs) and non-radiative silicon nitride (SiN_x) films. The band at 2360 cm^{-1} corresponds to the CO_2 molecule in the operating environment.

The gold nanoparticles (AuNPs) and silicon quantum dots (SiQDs) thin films were obtained using deposition conditions previously studied, which gave rise to average particle sizes of 2.9 nm and 3.1 nm, respectively [11,19]. It can be observed in both samples from HRTEM images (Figure 3a,b) a uniform distribution of particles throughout the whole surface and quasi-spherical shape. Likewise, the cover surface of gold nanoparticles obtained by HRTEM micrographs was 18.12% and its plasmonic resonance location was found at about 538 nm (inset of Figure 2a).

Cross-sectional SEM images of the PR2 (p-Si/SiN$_x$/SiQDs) and P2 (p-Si/AuNPs/SiN$_x$/SiQDs) structures are depicted in Figure 4a,b, respectively. From these images, the ZnO-Al transparent conductive contact can be identified at the top of the structures. It is worth noting that the SiN$_x$ and SiQDs layers are not distinguishable by this technique, in both samples and that the AuNPs in the P2 structure are clearly located at the interface between the silicon nitride and the silicon substrate. The XPS depth profiles of the PR2 and P2 samples from silicon nitride to silicon substrate are the inset of Figure 4a,b, respectively. These profiles show atomic concentrations of Si 2p, N 1s, and Cl 2p in the silicon nitride films, as well as no diffusion of them to the silicon substrate. Additionally, it is possible to identify that Au atoms do not diffuse to the bulk of the SiQDs layer. A content of oxygen is observed at the surface of these samples, which has been attributed to post-deposition reactions, occurred when the films were exposed to ambience [36,37].

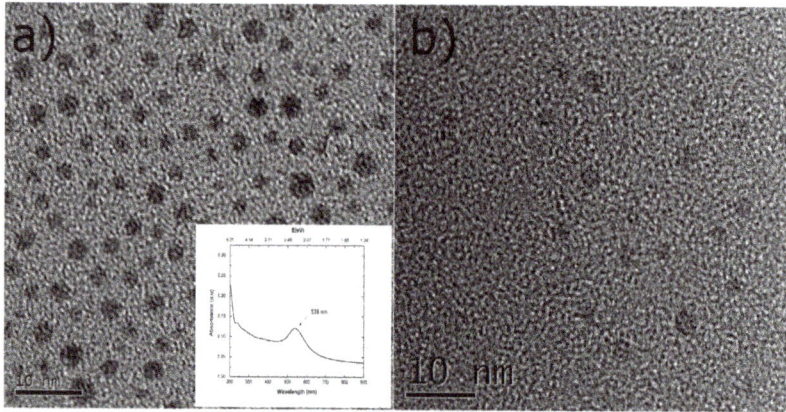

Figure 3. HRTEM images of gold nanoparticles (AuNPs) (**a**) and SiQDs (**b**) films. The average size and superficial density of AuNPs were 2.9 nm and 2.52×10^{12} particles/cm^2, respectively; meanwhile, for silicon nanoparticles, they were 3.1 nm and 6.04×10^{12} particles/cm^2, respectively [11,19]. The cover surface of gold nanoparticles was 18.12%, and its plasmonic resonance location was found at 538 nm (inset of Figure 3a).

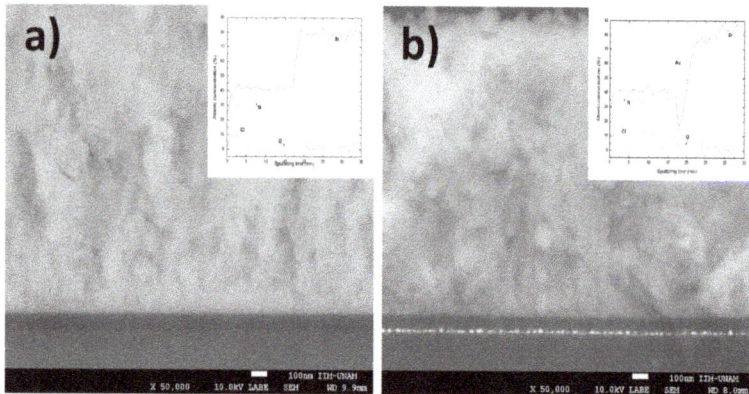

Figure 4. Cross-sectional views by SEM of the (**a**) PR2 (p-Si/SiN$_x$/SiQDs/ZnO-Al) and (**b**) P2 (p-Si/AuNPs/SiN$_x$/SiQDs/ZnO-Al) structures, respectively. The different silicon nitride layers (SiN$_x$ and SiQDs) are not distinguishable by this microscopy technique. The depth profiles of these samples from top silicon nitride to silicon substrate are inset of each figure.

3.2. Electroluminescence

The electroluminescence spectra of the four structures were obtained only in forward bias (considered when the cathode electrode is on the ZnO-Al) when applying voltages greater than 10 V, as can be seen from Figure 5. From this image, all the structures show an increment of the electroluminescent emission with increasing voltage at room temperature and a maximum intensity peak centered at around 600 nm. Also, it can be observed the influence of gold nanoparticles on the EL turn-on voltages of the fabricated structures, since for the PR1 and P1 samples (without the SiN_x layer), the EL turn-on voltages are 18 V and 14 V, respectively, i.e., lower for the sample with AuNPs. For these samples, voltage steps of 2 V are required to observe increased emission intensity. Likewise, the EL turn-on voltage is also lower for the P2 sample (11 V) with gold nanoparticles when compared with the reference PR2 (14 V) sample without them. Increased EL intensity is obtained using voltage steps of 1 V and 2 V for these samples, respectively.

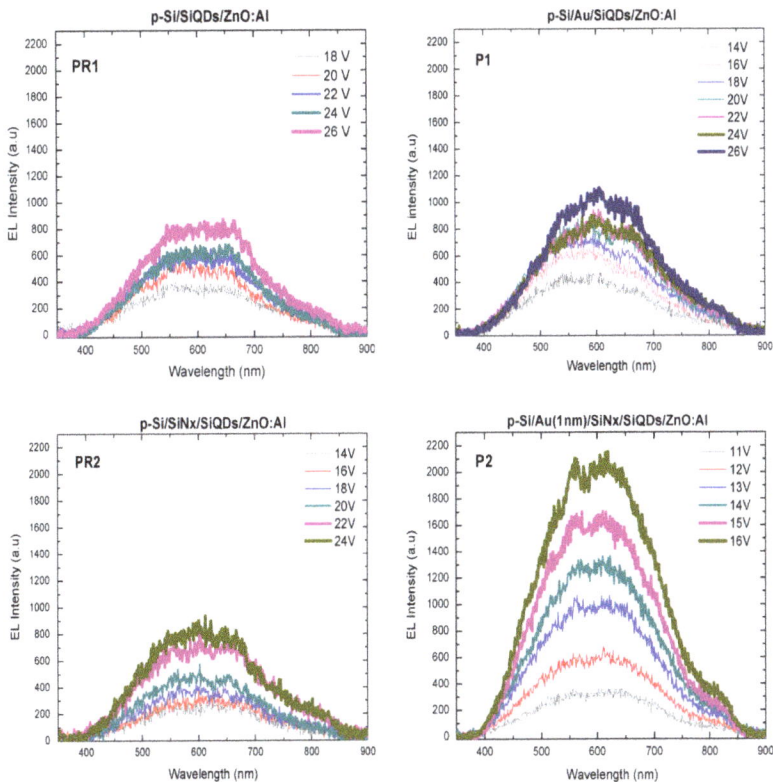

Figure 5. Electroluminescent (EL) spectra of the PR1 (p-Si/SiQDs/ZnO-Al), P1 (p-Si/AuNPs/ SiQDs/ZnO-Al), PR2 (p-Si/SiN$_x$/SiQDs/ZnO-Al) and P2 (p-Si/AuNPs/SiN$_x$/SiQDs/ZnO-Al) samples under forward bias.

The integrated electroluminescence intensity against the injected current of each sample is depicted in Figure 6. At first glance, a similar trend is observed for the P1 and PR1 samples; however, a slight EL enhancement (considered as the ratio of integrated EL intensity of devices with gold nanoparticles and reference devices) of 1.14 is found for these samples at about the same current of 23.5 mA. Moreover, the current injection is higher in the P1 sample when compared to the PR1 sample at the same applied voltage. The structures with the thin SiN_x layer (P2 and PR2) show an EL enhancement of 4.7 at ~27.7 mA. For these samples, the current injection is also higher for the sample with gold nanoparticles from 15 V. These results suggest an improvement of the external quantum efficiency in those samples using AuNPs.

In our previous paper, the PL emission of the four structures fabricated in this work had a Gaussian-like shape with a broad band centered at about 505 nm [19], manly attributed to quantum confinement effect in SiQDs. Since their corresponding EL emissions peaks were red shifted at ~600 nm, it is difficult to elucidate if the origin of EL is the same as that of PL, as other mechanisms with lower radiative transition probability as defects in the matrix or interface states in SiQDs (which produce radiative events at higher wavelengths) [27,38,39] could give rise to the observed EL emission.

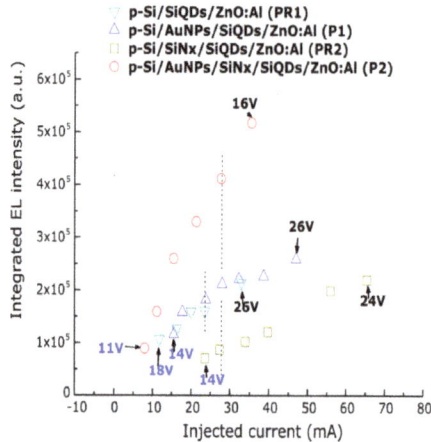

Figure 6. Integrated EL intensity vs injected current (mA) of the samples P1, PR1, P2, and PR2. A maximum EL enhancement (considered as the ratio of integrated EL intensity of devices with gold nanoparticles and reference devices) of 4.7 at ~27.7 mA is found for the P2 sample when compared to the reference PR2 sample.

The EL spectral enhancement factor, defined as the ratio of EL intensities of samples with gold nanoparticles and their references ones (IP1(λ)/IPR1(λ) and IP2(λ)/IPR2(λ)) is shown in Figure 7 for a defined injected current. For the IP1/IPR1 ratio at 23.5 mA, an almost constant line slightly above one is observed with some peaks at the beginning and at the end of the plot, probably due to the noise interfering with the measured signal at low emission intensities. This enhancement factor evaluated through a wavelength range from 430 to 815 nm has a maximum value of 1.45 at about 466 nm, which could indicate that the presence of gold nanoparticles could help to distribute carriers to the luminescent centers of the SiQDs film more efficiently in the P1 sample. On the other hand, the IP2/IPR2 ratio at 27.7 mA through the same wavelength range shows a maximum EL enhancement factor of 7 at about 510 nm, and 5.4 at about the maximum EL intensity peak (~600 nm). It is worth noting that the higher EL enhancement factor in these structures is close to the absorption peak of gold nanoparticles at 538 nm (inset of Figure 2a), which could suggest a resonant coupling between the silicon quantum dots emission and the oscillations in noble metal nanoparticles in the P2 sample [24]. Moreover, as the maximum IP1/IPR1 ratio was found to be lower than the maximum IP2/IPR2 ratio, in spite of the presence of metal nanoparticles in the P1 sample, it is important to take into account the role of the thin silicon nitride layer (SiN$_x$) separating the noble metal nanoparticles and the SiQDs film in the PR2 and P2 structures, which we will discuss later in this paper. The second factor of 5.4 is close to the EL enhancement of 4.7 earlier observed. Additionally, according to some authors [27,38–41], we observe shining spots on the top area of our devices, increasing in number as the current increased. Morales-Sánchez and Cabañas Tay et al. [40,41] explain the presence of these shining dots due to the formation of current paths connecting the top and bottom electrodes in silicon-rich oxide-based devices. Considering this, charge carriers under forward bias in our structures would create these paths hindered by the barriers of the bulk matrix and produced during their transport radiative and non-radiative transitions. An image of the luminescent dots obtained in the PR2 sample at 0.69 A/cm^2 is shown inset of Figure 7.

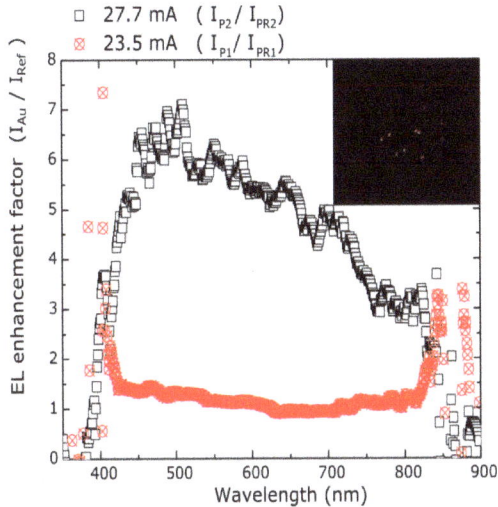

Figure 7. EL enhancement factor (defined as the ratio of EL intensities of samples with gold nanoparticles and their references ones (IP1(λ)/IPR1(λ) and IP2(λ)/IPR2(λ)). The IP1(λ)/IPR1(λ) ratio evaluated through a wavelength range from 430 to 815 nm at 23.5 mA, an almost constant line slightly above one is observed. The IP2(λ)/IPR2(λ) ratio at 27.7 mA shows a maximum EL enhancement factor of 7 at about 510 nm.

3.3. J-E Characteristics

Current density vs electric field (J–E) plots of the four structures studied in this work are depicted in Figure 8a,b. The P1 and PR1 samples have threshold electric fields of ~0.65 MV/cm and ~1.4 MV/cm, respectively. Just after current conduction is established, very similar slopes are observed in these structures (Figure 8a). It is possible to observe a lower threshold voltage in the P1 sample; this could be due to increased injection of carriers by the presence of gold nanoparticles. In addition, it is possible that the electrons from the gate could be drifted to the lower interface of the structure by areas of a higher electric field. Once there, the electrons can tunnel to the silicon nitride matrix, and to a lesser extent, to the SiQDs. The latter is possible in our device, since it has been studied that rms roughness as low as some nanometers at the bottom interface of a capacitor can be related to the increased leakage current through it, due to the enhanced local electric field at the protrusions [42,43]. Even though the morphology of gold nanoparticles could not be clearly detected by the AFM (Atomic Force Microscopy) technique, its average diameter size (estimated by TEM images of ~2.9 nm) suggests that this mechanism could be responsible for the lower threshold voltage and higher injection current at the same applied electric field in the P1 sample than PR1 sample. We could expect this effect to be less important at the top interface of the device, as these metal nanoparticles were not shown to increase, significantly, the surface roughness of the SiQDs layer with regard to the SiQDs layer of the reference device. Additionally, from previous results [19], no PL enhancement nor quenching of the P1 and PR1 structures was found, and we could indirectly assume that the reflectance due to the AuNPs is inappreciable. Therefore, the slight EL enhancement observed in these structures at about the same injected current should be due to a more efficient carrier injection to the luminescent centers of the active film in the P1 sample, and as a result, a higher number of conductive paths.

P2 and PR2 samples have threshold electric fields of ~1.16 MV/cm and ~0.68 MV/cm, respectively, i.e., lower for the reference structure (Figure 8b). For the PR2 sample, it is possible to observe current density under reverse bias, and higher forward current injection than in P2 sample, from low electric fields up to 1.75 MV/cm, after which the current increases faster in the P2 sample under a larger bias condition. The characteristics found in the PR2 sample could have their origin in the intrinsic properties of the silicon nitride films with different chemical composition. The presence of the very thin SiN_x layer (10 nm) with higher band gap than the SiQDs layer (according to the Tauc plots obtained) could increase the number of accumulated holes under forward bias in the p-Si surface at the bottom interface, and give rise to abrupt injection of carriers from the silicon substrate by tunneling towards the SiQDs layer. Moreover, the low refractive index of the SiN_x film indicates a high number of voids in the film, which would promote leakage current even at low electric fields. Since the integrated EL intensity in the PR2 sample was the lowest observed in the fabricated set of samples, we could deduce an inefficient carrier transport to the luminescent centers in the bulk of the active film, in spite of higher current passing through it from low electric fields.

On the other hand, the higher threshold voltage in the P2 sample may be due to a screening effect of the applied electric field by gold nanoparticles at the bottom interface of the structure [44]. This effect is not seen in the P1 sample (p-Si/AuNPs/SiQDs/ZnO-Al) where the electric field is higher, in a range from 0.74 MV/cm up to 1.49 MV/cm, than in P2 sample. One possible explanation is that silicon nanoparticles growing from the substrate in the P1 structure may work as conduction points of carriers and interact with the previously deposited AuNPs distributed throughout the substrate surface, making the p-Si/AuNPs/SiQDs interface inhomogeneous.

This direct interaction between silicon and gold nanoparticles is avoided by the low silicon content SiN_x layer in the P2 sample. Therefore, the layered configuration of the P2 sample, as a whole, should be responsible for its higher integrated EL intensity when compared with any of the other fabricated structures. The screening effect of the electric field observed in this sample could promote overlapping of the electron and hole population in regions far from the surface of the SiQDs film where the radiative recombination generally occurs [45]. Likewise, an almost five-fold enhancement of integrated EL intensity and the maximum EL enhancement factor IP2/IPR2 at about 510 nm could suggest an increment of the internal quantum efficiency by coupling of the local field near the surface of metal nanoparticles, and emission of the active layer at a distance defined by the SiN_x thickness (10 nm).

Since the integrated EL enhancement is larger than the integrated PL enhancement (~2) in the P2 sample [19] with regard to its reference one, it is possible that more than one mechanism is involved in the increase of electroluminescence observed, though a more detailed study is required to throw more light on the issue. Additionally, it was observed high reproducibility of the electrical characteristics of samples with gold nanoparticles, which is shown for three different devices (named D1, D2, D3), using the P1 and P2 configurations in Figure 8c,d, respectively. A double asterisk denotes the P1 and P2 samples in each figure labeled as the D3 device. The good reproducibility of the electrical and optical properties of these structures makes it possible its use in high-reliability applications.

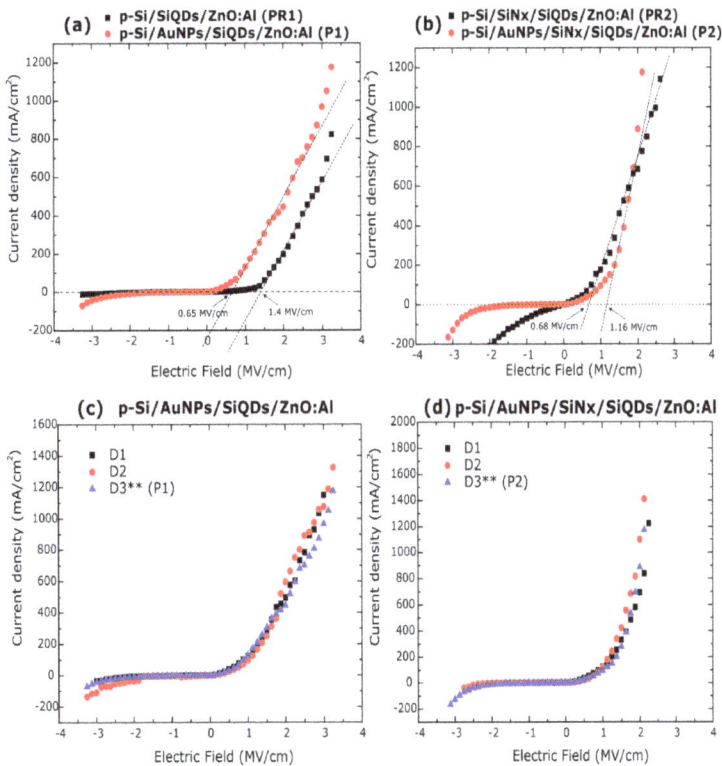

Figure 8. Current density-electric field characteristics of the (**a**) P1 and PR1 and (**b**) P2 and PR2 samples. Current density-electric field plots of three different devices of the P1 (**c**) and P2 (**d**) structures; a double asterisk denotes the P1 and P2 samples in each graph corresponding to the device D3.

4. Conclusions

We fabricated light emitting devices based on silicon nitride films layered structures that explored the role of AuNPs in the vicinity of a SiQDs active film. We observed improved carrier injection in samples using gold nanoparticles under determined bias condition, as well as lower electroluminescent (EL) turn-on voltages. Also, an EL enhancement was found in these samples when compared with their reference ones without noble metal particles, being higher for that sample using a non-radiative SiN$_x$ layer of 10 nm between the AuNPs and the SiQDs film (P2 sample). The almost five times higher integrated electroluminescence observed in this sample can be explained by considering the intrinsic physical properties of the different layers of nitride and metal that make up the device, such as band gap and roughness. However, it is also possible that the observed EL enhanced emission in this multilayer sample could be originated by the plasmonic coupling between AuNPs and the radiative SiQDs film.

Acknowledgments: All authors would like to acknowledge the technical assistance of Juan Manuel García León in different experimental aspects of this paper. Authors also acknowledge the support provided by Lázaro Huerta Arcos and M. A. Canseco in the use and interpretation of XPS and FTIR techniques respectively. The first author is grateful to the Dirección General de Asuntos del Personal Académico (DGAPA-UNAM) for the granted Postdoctoral Fellowship. This research work was financially supported by projects: (a) Investigación Científica Básica SEP—CONACYT 2016, project number: 283492 and (b) PAPIIT-UNAM, project number: IA102718, which are under the technical administration of Arturo Rodríguez-Gómez.

Author Contributions: Regarding the conception of the work and the writing of the article, Arturo Rodríguez-Gómez, Juan Carlos Alonso-Huitrón and Ana Luz Muñoz-Rosas worked in same amounts. Meanwhile, most of the experimental section, i.e., materials depositions and characterizations, were carried out by Ana Luz Muñoz-Rosas.

Conflicts of Interest: The authors declare no conflict of interest.

References

1. Abeles, B.; Tiedje, T. Amorphous Semiconductor Superlattices. *Phys. Rev. Lett.* **1983**, *51*, 2003–2006. [CrossRef]

2. DiMaria, D.J.; Kirtley, J.R.; Pakulis, E.J.; Dong, D.W.; Kuan, T.S.; Pesavento, F.L.; Theis, T.N.; Cutro, J.A.; Brorson, S.D. Electroluminescence studies in silicon dioxide films containing tiny silicon islands. *J. Appl. Phys.* **1984**, *56*, 401–416. [CrossRef]

3. Canham, L.T. Silicon quantum wire array fabrication by electrochemical and chemical dissolution of wafers. *Appl. Phys. Lett.* **1990**, *57*, 1046–1048. [CrossRef]

4. Pavesi, L.; Gaburro, Z.; Negro, L.D.; Bettotti, P.; Prakash, G.V.; Cazzanelli, M.; Oton, C.J. Nanostructured silicon as a photonic material. *Opt. Lasers Eng.* **2003**, *39*, 345–368. [CrossRef]

5. Streshinsky, M.; Ding, R.; Liu, Y.; Novack, A.; Galland, C.; Lim, A.E.-J.; Guo-Qiang Lo, P.; Baehr-Jones, T.; Hochberg, M. The Road to Affordable, Large-Scale Silicon Photonics. *Opt. Photonics News* **2013**, *24*, 32. [CrossRef]

6. Rickman, A. The commercialization of silicon photonics. *Nat. Photonics* **2014**, *8*, 579–582. [CrossRef]

7. Priolo, F.; Gregorkiewicz, T.; Galli, M.; Krauss, T.F. Silicon nanostructures for photonics and photovoltaics. *Nat. Nanotechnol.* **2014**, *9*, 19–32. [CrossRef] [PubMed]

8. Park, N.-M.; Choi, C.-J.; Seong, T.-Y.; Park, S.-J. Quantum Confinement in Amorphous Silicon Quantum Dots Embedded in Silicon Nitride. *Phys. Rev. Lett.* **2001**, *86*, 1355–1357. [CrossRef] [PubMed]

9. Kim, T.-W.; Cho, C.-H.; Kim, B.-H.; Park, S.-J. Quantum confinement effect in crystalline silicon quantum dots in silicon nitride grown using SiH$_4$ and NH$_3$. *Appl. Phys. Lett.* **2006**, *88*, 123102. [CrossRef]

10. Kim, T.; Park, N.; Kim, K.; Yong, G. Quantum confinement effect of silicon nanocrystals in situ grown in silicon nitride films. *Appl. Phys. Lett.* **2004**, *85*, 5355–5357. [CrossRef]

11. Rodriguez, A.; Arenas, J.; Alonso, J.C. Photoluminescence mechanisms in silicon quantum dots embedded in nanometric chlorinated-silicon nitride films. *J. Lumin.* **2012**, *132*, 2385–2389. [CrossRef]

12. Das, D.; Samanta, A. Quantum size effects on the optical properties of nc-Si QDs embedded in an a-SiO$_x$ matrix synthesized by spontaneous plasma processing. *Phys. Chem. Chem. Phys.* **2015**, *17*, 5063–5071. [CrossRef] [PubMed]

13. Alonso, J.C.; Pulgarín, F.A.; Monroy, B.M.; Benami, A.; Bizarro, M.; Ortiz, A. Visible electroluminescence from silicon nanoclusters embedded in chlorinated silicon nitride thin films. *Thin Solid Films* **2010**, *518*, 3891–3893. [CrossRef]

14. Biteen, J.S.; Pacifici, D.; Lewis, N.S.; Atwater, H.A. Enhanced Radiative Emission Rate and Quantum Efficiency in Coupled Silicon Nanocrystal-Nanostructured Gold Emitters. *Nano Lett.* **2005**, *5*, 1768–1773. [CrossRef] [PubMed]

15. Biteen, J.S.; Lewis, N.S.; Atwater, H.A.; Mertens, H.; Polman, A. Spectral tuning of plasmon-enhanced silicon quantum dot luminescence. *Appl. Phys. Lett.* **2006**, *88*, 131109. [CrossRef]

16. Mertens, H.; Biteen, J.S.; Atwater, H.A.; Polman, A. Polarization-Selective Plasmon-Enhanced Silicon Quantum-Dot Luminescence. *Nano Lett.* **2006**, *6*, 2622–2625. [CrossRef] [PubMed]

17. Benami, A.; López-Suárez, A.; Rodríguez-Fernández, L.; Crespo-Sosa, A.; Cheang-Wong, J.C.; Reyes-Esqueda, J.A.; Oliver, A. Enhancement and quenching of photoluminescence from silicon quantum dots by silver nanoparticles in a totally integrated configuration. *AIP Adv.* **2012**, *2*, 012193. [CrossRef]

18. Benami, A.; El Hassouani, Y.; Oliver, A.; Lopez-Suarez, A. Effect of Silver Nanoparticles on the Photoluminescence of Silicon Nanocrystals. *Spectrosc. Lett.* **2014**, *47*, 411–414. [CrossRef]

19. Muñoz-Rosas, A.L.; Rodríguez-Gómez, A.; Arenas-Alatorre, J.A.; Alonso-Huitrón, J.C. Photoluminescence enhancement from silicon quantum dots located in the vicinity of a monolayer of gold nanoparticles. *RSC Adv.* **2015**, *5*, 92923–92931. [CrossRef]

20. Biteen, J.S.; Sweatlock, L.A.; Mertens, H.; Lewis, N.S.; Polman, A.; Atwater, H.A. Plasmon-Enhanced Photoluminescence of Silicon Quantum Dots: Simulation and Experiment. *J. Phys. Chem. C* **2007**, *111*, 13372–13377. [CrossRef]

21. Inoue, A.; Fujii, M.; Sugimoto, H.; Imakita, K. Surface Plasmon-Enhanced Luminescence of Silicon Quantum Dots in Gold Nanoparticle Composites. *J. Phys. Chem. C* **2015**, *119*, 25108–25113. [CrossRef]

22. Yashima, S.; Sugimoto, H.; Takashina, H.; Fujii, M. Fluorescence Enhancement and Spectral Shaping of Silicon Quantum Dot Monolayer by Plasmonic Gap Resonances. *J. Phys. Chem. C* **2016**, *120*, 28795–28801. [CrossRef]

23. Inoue, A.; Sugimoto, H.; Fujii, M. Photoluminescence Enhancement of Silicon Quantum Dot Monolayer by Double Resonance Plasmonic Substrate. *J. Phys. Chem. C* **2017**, *121*, 11609–11615. [CrossRef]

24. Kim, B.-H.; Cho, C.-H.; Mun, J.-S.; Kwon, M.-K.; Park, T.-Y.; Kim, J.S.; Byeon, C.C.; Lee, J.; Park, S.-J. Enhancement of the External Quantum Efficiency of a Silicon Quantum Dot Light-Emitting Diode by Localized Surface Plasmons. *Adv. Mater.* **2008**, *20*, 3100–3104. [CrossRef]

25. Li, D.; Wang, F.; Ren, C.; Yang, D. Improved electroluminescence from silicon nitride light emitting devices by localized surface plasmons. *Opt. Mater. Express* **2012**, *2*, 872. [CrossRef]

26. Li, D.; Wang, F.; Yang, D. Evolution of electroluminescence from silicon nitride light-emitting devices via nanostructural silver. *Nanoscale* **2013**, *5*, 3435–3440. [CrossRef] [PubMed]

27. Kim, B.H.; Davis, R.F.; Cho, C.-H.; Park, S.-J. Enhanced performance of silicon quantum dot light-emitting diodes grown on nanoroughened silicon substrate. *Appl. Phys. Lett.* **2009**, *95*, 073113. [CrossRef]

28. Li, W.; Wang, S.; Hu, M.; He, S.; Ge, P.; Wang, J.; Guo, Y.Y.; Zhaowei, L. Enhancement of electroluminescence from embedded Si quantum dots/SiO_2 multilayers film by localized-surface-plasmon and surface roughening. *Sci. Rep.* **2015**, *5*, 11881. [CrossRef] [PubMed]

29. Rodríguez-Gómez, A.; Moreno-Rios, M.; García-García, R.; Pérez-Martínez, A.L.; Reyes-Gasga, J. Role of the substrate on the growth of silicon quantum dots embedded in silicon nitride thin films. *Mater. Chem. Phys.* **2018**, *208*, 61–67. [CrossRef]

30. Tauc, J. Optical properties and electronic structure of amorphous Ge and Si. *Mater. Res. Bull.* **1968**, *3*, 37–46. [CrossRef]

31. Robertson, J. Electronic structure of silicon nitride. *Philos. Mag. Part B* **1991**, *63*, 47–77. [CrossRef]

32. Aspnes, D.E.; Studna, A.A. Dielectric functions and optical parameters of Si, Ge, GaP, GaAs, GaSb, InP, InAs, and InSb from 1.5 to 6.0 eV. *Phys. Rev. B* **1983**, *27*, 985–1009. [CrossRef]

33. Anutgan, M.; Anutgan, T.A.; Atilgan, I.; Katircioglu, B. Photoluminescence analyses of hydrogenated amorphous silicon nitride thin films. *J. Lumin.* **2011**, *131*, 1305–1311. [CrossRef]

34. Ay, F.; Aydinli, A. Comparative investigation of hydrogen bonding in silicon based PECVD grown dielectrics for optical waveguides. *Opt. Mater.* **2004**, *26*, 33–46. [CrossRef]

35. Jiang, X.; Ma, Z.; Xu, J.; Chen, K.; Xu, L.; Li, W.; Huang, X.; Feng, D. a-SiN$_x$:H-based ultra-low power resistive random access memory with tunable Si dangling bond conduction paths. *Sci. Rep.* **2015**, *5*, 15762. [CrossRef] [PubMed]

36. Serrano-Núñez, M.A.; Rodríguez-Gómez, A.; Escobar-Alarcón, L.; Alonso-Huitrón, J.C. Combined study of the effect of deposition temperature and post-deposition annealing on the photoluminescence of silicon quantum dots embedded in chlorinated silicon nitride thin films. *RSC Adv.* **2016**, *6*, 77440–77451. [CrossRef]

37. Rodríguez-Gómez, A.; Escobar-Alarcón, L.; Serna, R.; Cabello, F.; Haro-Poniatowski, E.; García-Valenzuela, A.; Alonso, J.C. Modeling of the refractive index and composition of luminescent nanometric chlorinated-silicon nitride films with embedded Si-quantum dots. *J. Appl. Phys.* **2016**, *120*, 145305. [CrossRef]

38. Wang, M.; Xie, M.; Ferraioli, L.; Yuan, Z.; Li, D.; Yang, D.; Pavesi, L. Light emission properties and mechanism of low-temperature prepared amorphous SiNX films. I. Room-temperature band tail states photoluminescence. *J. Appl. Phys.* **2008**, *104*, 083504. [CrossRef]

39. Wang, M.; Huang, J.; Yuan, Z.; Anopchenko, A.; Li, D.; Yang, D.; Pavesi, L. Light emission properties and mechanism of low-temperature prepared amorphous SiN$_X$ films. II. Defect states electroluminescence. *J. Appl. Phys.* **2008**, *104*, 083505. [CrossRef]

40. Morales-Sánchez, A.; Domínguez, C.; Barreto, J.; Aceves-Mijares, M.; Licea-Jiménez, L.; Luna-López, J.A.; Carrillo, J. Floating substrate luminescence from silicon rich oxide metal-oxide-semiconductor devices. *Thin Solid Films* **2013**, *531*, 442–445. [CrossRef]

41. Cabañas-Tay, S.A.; Palacios-Huerta, L.; Luna-López, J.A.; Aceves-Mijares, M.; Alcántara-Iniesta, S.; Pérez-García, S.A.; Morales-Sánchez, A. Analysis of the luminescent centers in silicon rich silicon nitride light-emitting capacitors. *Semicond. Sci. Technol.* **2015**, *30*, 065009. [CrossRef]

42. Gaillard, N.; Pinzelli, L.; Gros-Jean, M.; Bsiesy, A. In situ electric field simulation in metal/insulator/metal capacitors. *Appl. Phys. Lett.* **2006**, *89*, 89–92. [CrossRef]

43. Lopes, M.C.V. Si-SiO$_2$ Electronic Interface Roughness as a Consequence of Si-SiO$_2$ Topographic Interface Roughness. *J. Electrochem. Soc.* **1996**, *143*, 1021–1025. [CrossRef]

44. Regan, W.; Byrnes, S.; Gannett, W.; Ergen, O.; Vazquez-Mena, O.; Wang, F.; Zettl, A. Screening-Engineered Field-Effect Solar Cells. *Nano Lett.* **2012**, *12*, 4300–4304. [CrossRef] [PubMed]

45. Oguro, T.; Koyama, H.; Ozaki, T.; Koshida, N. Mechanism of the visible electroluminescence from metal/porous silicon/n-Si devices. *J. Appl. Phys.* **1997**, *81*, 1407–1412. [CrossRef]

![nanomaterials logo] *nanomaterials*

MDPI

Article

On-Demand CMOS-Compatible Fabrication of Ultrathin Self-Aligned SiC Nanowire Arrays

Natasha Tabassum [1], **Mounika Kotha** [1], **Vidya Kaushik** [1], **Brian Ford** [1], **Sonal Dey** [1],
Edward Crawford [2], **Vasileios Nikas** [1] and **Spyros Gallis** [1,*]

[1] Colleges of Nanoscale Sciences and Engineering, SUNY Polytechnic Institute (SUNY Poly), Albany, NY 12203, USA; ntabassum@sunypoly.edu (N.T.); mkotha@sunypoly.edu (M.K.); vkaushik@sunypoly.edu (V.K.); Brford@sunypoly.edu (B.F.); sonal_dey@avs.org (S.D.); vnikas@sunypoly.edu (V.N.)
[2] GLOBALFOUNDRIES Corp., East Fishkill, NY 12533, USA; Edward.Crawford@globalfoundries.com
[*] Correspondence: sgalis@sunypoly.edu; Tel.: +1-518-956-7048

Received: 20 October 2018; Accepted: 3 November 2018; Published: 5 November 2018

Abstract: The field of semiconductor nanowires (NWs) has become one of the most active and mature research areas. However, progress in this field has been limited, due to the difficulty in controlling the density, orientation, and placement of the individual NWs, parameters important for mass producing nanodevices. The work presented herein describes a novel nanosynthesis strategy for ultrathin self-aligned silicon carbide (SiC) NW arrays ($\leq$ 20 nm width, 130 nm height and 200–600 nm variable periodicity), with high quality (~2 Å surface roughness, ~2.4 eV optical bandgap) and reproducibility at predetermined locations, using fabrication protocols compatible with silicon microelectronics. Fourier transform infrared spectroscopy, X-ray photoelectron spectroscopy, ultraviolet-visible spectroscopic ellipsometry, atomic force microscopy, X-ray diffractometry, and transmission electron microscopy studies show nanosynthesis of high-quality polycrystalline cubic 3C-SiC materials (average 5 nm grain size) with tailored properties. An extension of the nanofabrication process is presented for integrating technologically important erbium ions as emission centers at telecom C-band wavelengths. This integration allows for deterministic positioning of the ions and engineering of the ions' spontaneous emission properties through the resulting NW-based photonic structures, both of which are critical to practical device fabrication for quantum information applications. This holistic approach can enable the development of new scalable SiC nanostructured materials for use in a plethora of emerging applications, such as NW-based sensing, single-photon sources, quantum LEDs, and quantum photonics.

Keywords: silicon carbide; ultrathin nanowires; nanofabrication; self-aligned nanowires; telecom wavelengths; quantum photonics

1. Introduction

As the field of semiconductor nanowires (NWs) has become one of the most active and relatively mature research areas, heightened interest in the synthesis, characterization, and applications of these NWs has become prevalent within the scientific community. The unique properties of ultrathin NWs, resulting from their reduced dimensionality coupled with their tunable properties and surface functionalization, make them promising for various applications in the field of electronics [1], optics [2], biological sciences [3–5], medical diagnosis [6], energy harvesting [7], and ultra-high nanosensing [8]. Furthermore, great efforts have been focused on the development of nanostructured materials that may be employed in emerging quantum applications, such as quantum imaging and sensing, and quantum photonics [9,10].

Top-down and bottom-up approaches are the two basic paradigms of nanofabrication. The top-down approach refers to the etching of bulk material to sculpt nanostructures. This has the primary advantage of direct assembly after processing, as the density and spatial location of the resulting nanostructure are defined by the process. However, top-down fabrication is heavily dependent on lithographical patterning of the desired features and thus, fabricating features below 20 nm becomes difficult and costly, and the composition of the resulting nanostructured materials is limited to the bulk materials from which they are fabricated.

The bottom-up approach allows for the fabrication of a wide range of materials by assembling the required subcomponents in an additive fashion. Common techniques in this paradigm include vapor-liquid-solid [11], template-assisted electrochemical deposition [12], and solution-based growth strategies [13]. The advantages of bottom-up techniques include the synthesis of a wide range of both inorganic and organic materials, synthesis of heterostructures, such as axial and radial core/shell structures, and the ability to dope in situ. The main limiting challenge commonly faced by this paradigm is the required deterministic assembly, which involves control over the density, orientation, spacing, and placement of the individual NWs, and their integration into large-scale arrays with high scalability and reproducibility [14].

Deterministic assembly of NW arrays is essential for the mass production of electronic nanodevices and the creation of practical nanoscale-based systems for the fabrication of functional interconnected nanosystems [14]. Furthermore, the realization of devices in the emerging field of quantum technologies requires innovative nanomaterial architectures, where the optical and quantum properties of emission centers can be deterministically engineered [10].

Silicon Carbide (SiC) is a silicon-based wide band-gap material, which exhibits strong mechanical properties and is chemically inert. These properties have led to the employment of SiC in a variety of applications due their stability within a multitude of environments: In high-temperature energy conversion devices and in chemically corrosive and high shock environments [15,16]. SiC-based field-effect transistors (FETs) have demonstrated operation for thousands of hours under high-temperature conditions [17] and SiC NWs have been used in hydrogen sensors [18]. Furthermore, owing to its high biocompatibility [19], SiC is used in the biomedical field for coating implants, as SiC nanostructures enhance cell proliferation and accelerate tissue reconstruction [20,21]. In the field of nanobiotechnology, SiC NW-FETs have demonstrated the ability to detect DNA hybridization [22]. Moreover, naturally occurring Si and C in SiC has almost negligible magnetic moment [23], which is a necessary requirement for hosting quantum emitters with reduced optical decoherence caused by nuclear and electronic spin fluctuation [24]. In that regard, SiC nanophotonic structures have been recognized as promising systems for several applications in quantum technologies [10].

Herein, the current investigators present an innovative and straightforward synthesis route for SiC NW arrays. This synthesis route allows for ultrathin self-aligned NWs to be fabricated without the use of a lithographic-pattern-transfer technique. This fabrication scheme overcomes obstacles faced by top-down and bottom-up approaches, which typically result in high surface defect density states, due to the dry-etch step or random orientation and size, non-specific positioning, or requires transfer to another substrate and subsequent fabrication steps. Ultrathin array NWs ($\leq$20 nm) are advantageous for NW array-based biosensors with high sensitivity, as the reduced size allows for full gating of the NW by the charged species of interest [8]. Ultrathin SiC NW allows for higher mechanical strength [25] and reduced bulk-defect density [26], which is particularly beneficial for the emission of color centers [27,28]. This synthesis strategy may serve as a common experimental platform to investigate multiple SiC NW-based emerging technologies, such as NW-based sensing, single photon sources, quantum LEDs and quantum photonics. To this end, we have extended the above mentioned fabrication process to host erbium ions in SiC NW arrays. The integration scheme allows us to control the locations of erbium ions in SiC NWs and to modify the spontaneous emission properties of these ions. Both are important components towards potential device applications in the emerging field of quantum photonics.

2. Materials and Methods

A simplified schematic representation with corresponding scanning electron microscope images (SEMs) of the growth strategy for ultrathin self-aligned SiC NW arrays is shown in Figure 1. First, ribbon arrays (250 × 250 μm^2) of hydrogen silsequioxane (HSQ) negative-tone resist were fabricated on clean Si (100) substrates using electron-beam lithography (EBL). The Si substrate was spin-coated with HSQ (6 wt.% HSQ in methyl isobutyl ketone (MIBK)), followed by a soft-bake at 80 °C for 4 min prior to exposure. The HSQ resist layer was exposed with 100 kV beam using VB300 EBL system (Vistec Electron Beam GmbH, Jena, Germany) or with a 50 kV beam using Voyager system (Raith GmbH, Dortmund, Germany). Exposed resist layer then developed in 2.38 wt.% of tetra methyl ammonium hydroxide (TMAH), yielding an HSQ ribbon array (Figure 1a). Electron-beam lithography was used for rapid prototyping while minimizing expenses; however, with the proper photomasks and processes, the fabrication scheme presented is seamlessly compatible with standard photolithography by using a thin oxide or nitride film instead of the resist layer (Figure 1a).

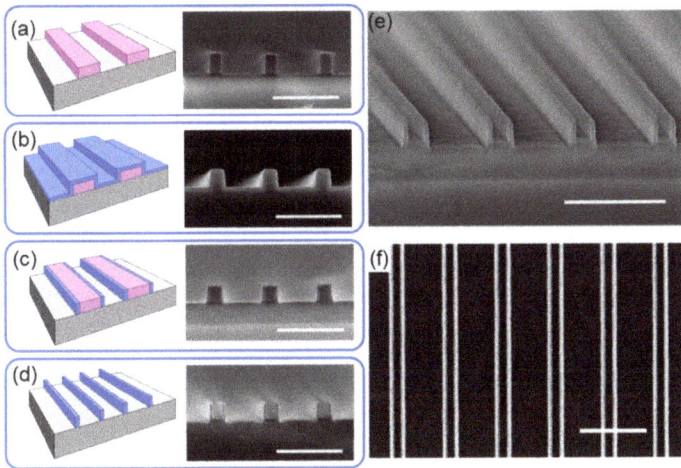

Figure 1. Nanofabrication of ultrathin self-aligned nanowires (NW) array. (a) Si wafer (gray) was spin-coated with hydrogen silsequioxane (HSQ) followed by exposure and development yielding a ribbon array (pink) with width and pitch ranging from 50 nm to 150 nm and from 200 nm to 600 nm, respectively; (b) ultrathin conformal silicon carbide (SiC) layer (blue) was deposited using thermal CVD; (c) ultrathin conformal SiC layer was etched open to expose the HSQ ribbon array using inductively coupled plasma reactive ion etching (ICP-RIE); (d) removal of the HSQ ribbon array was done by wet etch in buffered hydrofluoric acid (BHF), yielding SiC NW arrays with 20 nm critical dimension (width) NWs. Cross-section SEM images are shown after corresponding steps; (e) tilted cross-section and (f) top-down SEM image of the SiC NW array. Scale bar in all SEM images is 500 nm.

Nanowire synthesis was conducted using a well-controlled thermal chemical vapor deposition (CVD) process, which has been reported previously for the synthesis of silicon carbide (SiC) and silicon oxycarbide (SiC:O) [29–31]. For this work, ultrathin (10 to 40 nm) SiC or SiC:O was deposited onto the HSQ ribbon array, followed by a thermal anneal for 1 hour in forming gas (5% H$_2$, 95% Ar) at a temperature ranging from 900 to 1200 °C. The thickness of the SiC conformal layer synthesized onto the HSQ ribbon array (Figure 1b) defines the critical dimension (width) of the NWs, hence the NW width is solely dependent on the deposition process and not any lithographic transfer/post-material-synthesis etching.

The Si and C single-source oligomer used was CVD-742 (1,1,3,3-tetramethyl-1,3-disilacyclobutane, Starfire Systems) along with forming gas (5% H$_2$, 95% N$_2$) as a dilution gas. The SiC conformal layer was synthesized at 800 °C while system pressure was maintained at 1.0 Torr with a precursor flow rate fixed at 10 sccm. For optional oxygen doping (synthesis of SiC:O), synthesis parameters (temperature, pressure, precursor flow rate) were identical to the SiC synthesis with the exception of dilution gas and the use of a co-reactant, which were ultra-high purity Ar and O_2 respectively.

After synthesis of the conformal layer of SiC or SiC:O on the HSQ ribbon array, inductively coupled plasma reactive ion etching (ICP-RIE) was performed to remove the undesired material (ultrathin film between and on top of HSQ ribbons) and to expose the HSQ ribbon array for subsequent wet-etch removal (Figure 1c). ICP-RIE was performed at 7 mTorr using a ratio of 90:10 $CHF_3:O_2$ for 10–15 seconds. After etching, the NW array fabrication was completed upon removal of the HSQ ribbon array by dipping the sample in buffered hydrofluoric acid (BHF) for 5 min (Figure 1d). Representative cross-section and top-down SEM images of the resulting nanowire array structure are shown in Figure 1e,f.

For photoluminescence (PL) measurements, we used a home-built micro-PL (μPL) system–composed of an argon laser (model: Beamlock 2065, Spectra-Physics, Santa Clara, CA, USA), a dichroic mirror (DM), a 50× objective lens, a scanning nano-stage (1 nm resolution) for sample positioning, a fiber coupled FLSP920 spectrometer (Edinburgh Instruments, Livingston, UK) and an InGaAs detector.

3. Results and Discussion

3.1. Nanofabrication

Lithography parameters for the HSQ ribbon array were identified by performing a dose array study from 1200 to 1750 μC/cm^2 on a representative layout for the ribbon array with 100 nm wide lines with a pitch of 400 nm. A dose of 1400 μC/cm^2 was observed to best replicate the designed dimensions (Figure 2b). A lower dose of 1300 μC/cm^2 (Figure 2a) resulted in thinner ribbons indicating under-exposure and a higher dose at 1600 μC/cm^2 (Figure 2c) resulted in thicker ribbons with flared-out base indicating over-exposure. The optimal development time for the HSQ ribbon array was found by increasing the development time in 4-min increments. Shown in Figure 3c, 16 min was required for complete removal of the HSQ residue between ribbons. It is worth noting that the HSQ residue was removed more quickly at the edge of the ribbon array (Figure 3, left column) compared to the center (Figure 3, right column).

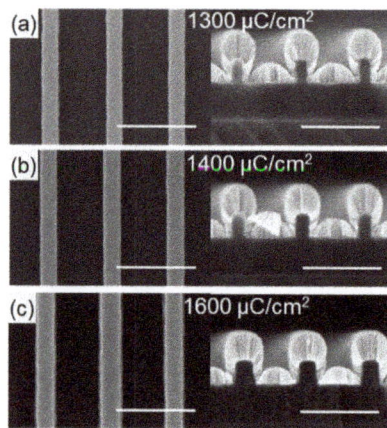

Figure 2. Dose array for HSQ ribbon array. Top-down (left) and cross-section (right, with metallization for imaging contrast) SEM images of approximately 100 nm wide HSQ ribbons after exposure at (a) 1300 μC/cm^2; (b) 1400 μC/cm^2 and (c) 1600 μC/cm^2. For all the SEM images shown, scale bar is 500 nm.

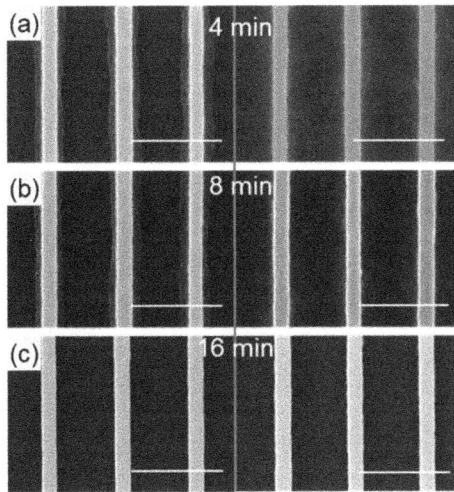

Figure 3. Development time for HSQ ribbon array. Comparison of development time at the edge (left column) and center (right column) of the HSQ ribbon array after (**a**) 4 min; (**b**) 8 min, and (**c**) 16 min of development time. Scale bar is 500 nm for all the SEM images shown.

By using the proposed nanofabrication approach, where the critical dimension of the NWs is defined by a well-controlled deposition process, production complexity can be reduced. Furthermore, the NW height can be controlled by changing the HSQ thickness with different parameters during spin-coating. Most importantly, this integration scheme can be material-invariant with proper etch and deposition techniques.

Flexibility of the growth strategy was explored by creating HSQ ribbon arrays with a pitch-to-ribbon-width ratio of 4:1. The pitch is denoted as P_1 and the ribbon width, which becomes sub pitch of the resulting NWs is denoted as P_2. Ribbon arrays were fabricated with dimensions of P_1:P_2-600:150, 500:125, 400:100, 300:75, and 200:50 (all numbers are in nm). A schematic depiction of the resulting structures is shown in Figure 4a with corresponding SEMs in Figure 4b–f. The critical dimension (width) and the spacing of the NWs can be modulated by adjusting the deposition time and customizing the lithography accordingly based on specific application requirements, such as increased spacing for subsequent processing.

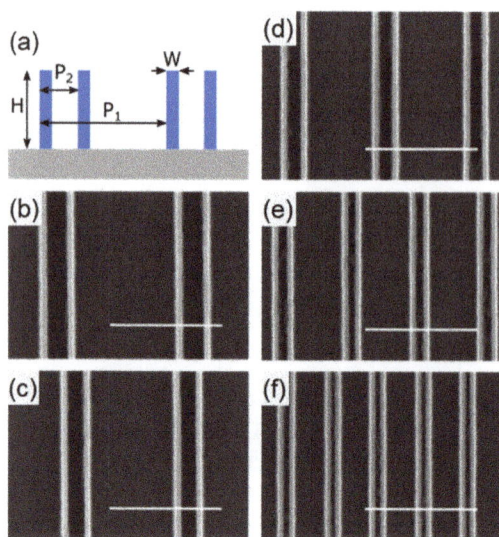

Figure 4. Modulation of NW array pitch (P₁) and sub-pitch (P₂). (a) Schematic representation of the resulting NW array structure with pitch (P_1) and sub-pitch (P_2), and NW height (H) and width (W) by modifying the layout of the HSQ ribbon array. Arrays of 20 nm (width, W) SiC NWs with a P_1 to P_2 ratio of 4:1 were fabricated. After nanofabrication, SEM images were collected for P_1:P_2; (b) 600:150; (c) 500:125; (d) 400:100; (e) 300:75; and (f) 200:50 (all numbers are in nm). Scale bars in all the SEM images are 500 nm.

3.2. Structural, Compositional, Optical, and Morphological Properties

Quality of the synthesized SiC nanomaterial, i.e., stoichiometry, crystal phase, surface roughness, defect density etc., are very important to investigate for any practical device application. Different characterization analyses, such as Fourier transform infrared spectroscopy (FTIR), X-ray photoelectron spectroscopy (XPS), ultraviolet visible spectroscopic ellipsometry (UV-VIS-SE), atomic force microscopy (AFM), high-resolution scanning transmission electron microscopy (HR-STEM) and X-ray diffractometry (XRD) were systematically carried out to assess and optimize the deposition and post-deposition process parameters towards achieving high-quality SiC nanowires.

FTIR spectroscopy showed a single strong absorption peak at ~760 cm^{-1} for all synthesized SiC, corresponding to Si–C stretching mode [32,33]. As shown in Figure 5a, upon increasing annealing temperature, T_A, three notable changes in the absorption spectra were observed: (1) A shift in peak position towards 800 cm^{-1} (Figure 5a), which corresponds to the stretching mode of Si–C bond in crystalline SiC [33,34], (2) substantial narrowing of the full width at half maximum (FWHM) (Figure 5b) and, (3) a change in line shape from Gaussian to Lorentzian. The line shape of the Si–C stretching mode of the as-deposited (AD) SiC was fitted with a Gaussian function of ~265 cm^{-1} FWHM, indicating a Gaussian distribution of bond lengths and angles, which characterizes the amorphous phase. Moreover, a transformation of the line shape to Lorentzian, corresponding to a more uniform environment of Si–C bonds, suggest the formation of crystalline SiC in the materials [35,36]. Upon annealing, the FWHM decreased to ~30 cm^{-1}, comparable to values for high-quality SiC [37].

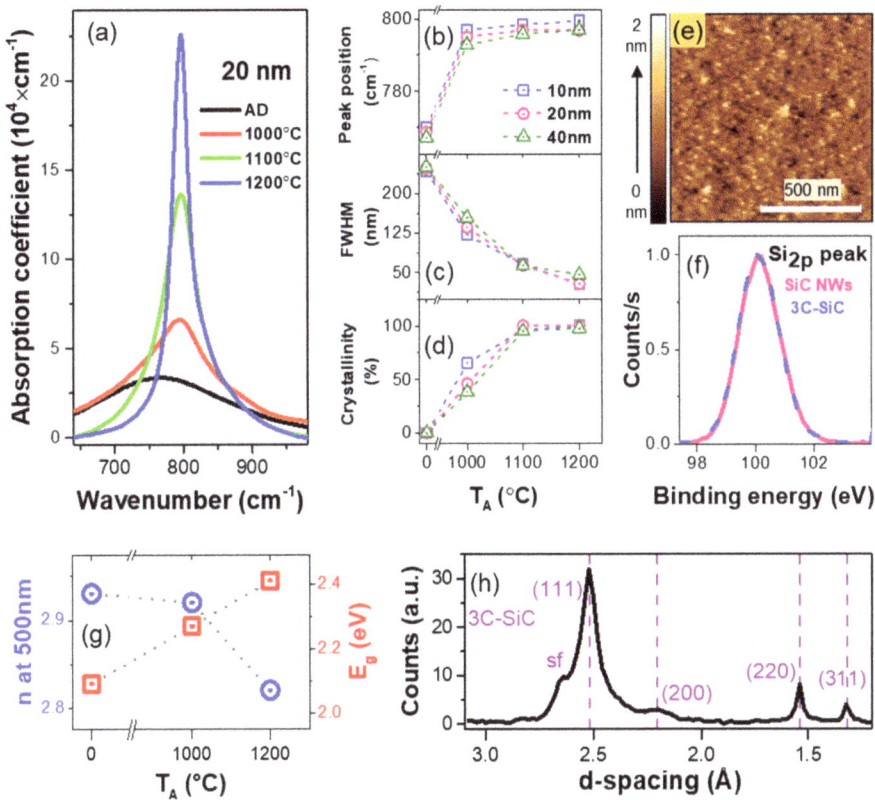

Figure 5. **Structural, compositional and optical analysis of the synthesized SiC.** (**a**) The Si–C stretching mode of the as-deposited (AD) and annealed 20-nm SiC at 1000, 1100, and 1200 °C in FG (5% H2, 95% Ar); (**b**) peak position of the Si–C stretching mode; (**c**) full width at half maximum (FWHM) of the Si–C stretching mode; and (**d**) crystalline fraction in materials as a function of annealing temperature, T_A. Error bars are not depicted as the errors are smaller than the symbol size; (**e**) representative atomic force microscopy (AFM) image of the surface of 20 nm SiC ultrathin film after 1100 °C anneal; (**f**) X-ray photoelectron spectroscopy (XPS) data of Si 2p peak from synthesized SiC and 3C-SiC control sample; (**g**) refractive index, n at 500 nm and Tauc optical gap, E_g vs. T_A for SiC, where errors are smaller than the symbol size; (**h**) X-ray diffractometry (XRD) pattern of 1200 °C-annealed SiC. The dashed lines correspond to the d-spacing values of (111), (200), (220) and (311) planes of 3C-SiC phase; sf, stacking faults.

Furthermore, to isolate the contribution of amorphous and crystalline phase, we deconvoluted the FTIR absorption spectra of the AD and annealed SiC materials into Gaussian (G) and Lorentzian (L) components, then the areas of each component were used to determine the crystalline fraction by calculating L/(L + G) (Figure 5d). For example, for the 10 and 20 nm ultrathin films, it was observed that annealing at 1100 °C for one hour resulted in complete crystallization as the absorption band could be fit using a single Lorentzian. Increasing the annealing temperature further only decreased the FWHM. The overall transformation of FTIR absorption spectra of Si–C mode with T_A suggests a high-quality and degree of crystalline environment of the synthesized SiC.

AFM was performed on SiC ultrathin films in order to assess the surface morphology. A representative AFM micrograph of a representative 1100 °C-annealed SiC sample is shown in

Figure 5e, where the mean surface roughness was determined to be ~2 Å, which is comparable to a polished Si wafer [38].

We performed compositional analysis of the annealed SiC via XPS. The ratio of silicon and carbon was found to be 50:50 within <1% experimental uncertainty. Additionally, strong overlap for the Si 2p binding energy between the annealed SiC and a 3C–SiC standard at ~100.3 eV confirms the chemical bonding of the synthesized materials and NWs are indeed Si–C (Figure 5f). The absence of XPS Si 2p peak around ~98.5 and ~103 eV, corresponding to Si–Si and Si–O bonds respectively [32], also ruled out the possibility of silicon or carbon nanocluster formation, as well as any oxidation.

The refractive index n, and Tauc optical gap E_g, of the SiC nanomaterials further elucidate the quality of synthesized materials. We extracted both parameters from UV-VIS-SE measurements (SOPRALAB, Semilab Co. Ltd., Budapest, Hungary). A decrease in refractive index upon annealing at higher temperatures was observed (Figure 5g), approaching the reference value for 3C–SiC (~2.7) [39]. E_g values were calculated using Tauc's law $\alpha E = B(E - E_g)^2$, where α is the absorption coefficient, B is the slope (which is inversely proportional to the band tail width), E is the photon energy, and E_g is the optical bandgap [29]. E_g increased with higher annealing temperatures, approaching the reference value of ~2.4 eV for 3C–SiC [40].

To determine the crystalline phase of the material, XRD analysis was done on different samples. The grazing incidence XRD (GIXRD) experiments were done at the Cornell High Energy Synchrotron Source (CHESS) using X-rays of energy 11.4 keV. The details of the experimental setup can be found elsewhere [41]. The vertical lines in the grazing incidence d-spacing map (GIDSM) confirmed the polycrystalline nature of the material. The GIDSM data were integrated to obtain an intensity versus d-spacing GIXRD plot for a 40 nm 1200 °C-annealed SiC (Figure 5h). The indexing was done with JCPDS #00-029-1129 (SiC, FCC, space group #216; a = 4.35890 Å). The peak positions align well with 3C–SiC phase. The d-spacing value at 2.52, 2.2, 1.54 and 1.32 (corelates to 2θ = 35.6°, 40.9°, 60.0°, and 70°) correspond to the (111), (200), (220), and (311) planes observed in 3C–SiC [42]. Predominant peak of (111) plane signifies that the 3C-SiC nanocrystals have a preferred orientation of growth. We used both Scherrer formula [34] and Williamson-Hall plot [43] to estimate the average grain sizes of the annealed SiC. From Scherrer formula, we found the grain size to be around 3–5 nm. The calculated grain size from Williamson-Hall plot's intercept was found to be approximately 4.5 nm. Peak broadening may be attributed to strain or size-confinement, and the shoulder at 2.64 Å may be due to stacking faults in SiC [33,43].

HR-STEM along with energy-dispersive spectroscopy (EDS) studies were also performed, as shown in Figure 6. The images showed that the NW is polycrystalline with small grains size of average ~5 nm (Figure 6b), confirming the calculated grain size from XRD. Indexing of the digital diffraction pattern was obtained by FFT (Fast Fourier Transform). FFT showed ring patterns with radii of 2.51 Å, 2.22 Å, 1.55 Å and 1.34 Å corresponding well to the inter-planar spacings of cubic 3C-SiC (Figure 6b). The TEM image and the discontinuity of the observed ring at certain angles suggest a preferential growth direction of the 3C-SiC grains in ultrathin NWs [44]. The spacing between the lattice fringes of a single grain was, on average, ~2.55 Å (Figure 6c), which is close to d-spacing, ~2.52 Å, of the (111) plane of 3C–SiC [45]. FFT on that single grain showed only two bright spots with ~2 × 2.51 Å distance, suggesting a single crystalline 3C–SiC. The EDS maps confirm the presence of Si and C in the NW with no oxidation.

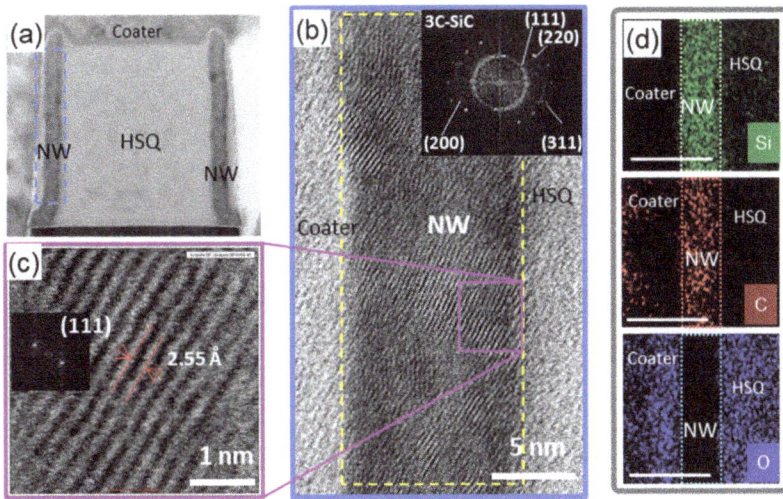

Figure 6. Transmission electron microscopy (TEM) analysis of SiC NW array with $P_1:P_2 = 400:100$ nm (**a**) Representative TEM image of a pair of 10 nm (width, W) SiC NWs; (**b**) High-resolution TEM image of a single nanowire. Inset: A fast Fourier transform (FFT) of the yellow framed area in (**b**), showing spotted rings with approximate radii of 2.5, 2.22, 1.55 and 1.34 Å corresponding to (111), (200), (220), (311) planes of 3C–SiC; (**c**) HRTEM (high-resolution scanning TEM) image from the purple framed area in (**b**) showing the (111) orientation, which is confirmed by the FFT in inset; (**d**) Elemental analyses of an area surrounding a single 10 nm NW.

3.3. Deterministic Ion Integration Into NW Arrays

We extended the growth strategy for synthesizing self-aligned SiC NW arrays and implemented it for the deterministic placement of erbium (Er) ions into the NW arrays (Figure 7). The proposed integration scheme is not specific to erbium ions, opening up a great potential for the use of SiC nanophotonic structures for applications in quantum information and quantum photonics [10,46]. The extended fabrication scheme begins after the ICP-RIE to expose the HSQ ribbon array and before the wet-etch step previously discussed (see Figure 7c). First, a ~200 nm thick sacrificial oxide was deposited using thermal CVD. The oxide was planarized using chemical mechanical planarization and was then recessed to expose the top of the NWs using ICP-RIE followed by a wet-etch in BHF for 10 seconds, as shown in Figure 7(d-1,2). Following the oxide recession, an encapsulation oxide was deposited targeting 15 nm thickness prior to ion implantation (Figure 7(d-3,4)). The encapsulation oxide (Figure 7(d-3,4)) serves two purposes: (1) It allows for tailoring the target implantation depth of the ions into the NWs, and (2) it protects the NW surface from ion implantation damage. Erbium ion implantation was performed using an Extrion 400 Ion Implanter, targeting 45 nm implantation depth. The doses of erbium ions were varied between 1×10^{13}–1×10^{14} cm^2, a typical range for characterizing emission behavior of ions minimizing inter-ionic interaction [47]. After the ion implantation, the samples were wet-etched in BHF for five minutes to remove the encapsulation oxide, sacrificial oxide, and HSQ ribbon array, and annealed at 900 °C for one hour in ultra-high purity Ar to optically activate the Er ions (Er^{3+}) based on our group's previous study [48].

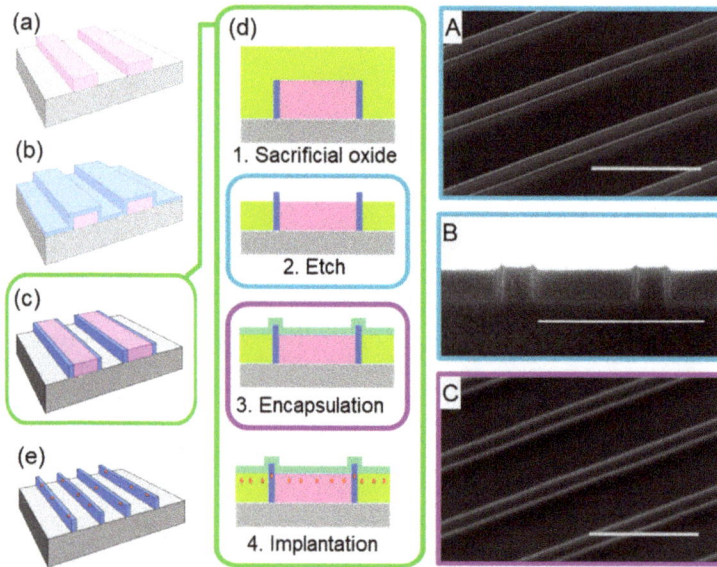

Figure 7. Integration scheme for controlled ion implantation into NW array. Schematics (a)–(c) correspond to Figure 1a–c; (d) schematic representation of the process steps involved in the ion implantation; green box continued from (c): (1) Thick sacrificial silicon oxide layer (light green) deposition. (2) Sacrificial oxide planarized and recessed below the tips of the NWs. Corresponding top-down SEM shown in **A** with cross-section in **B** (scale bar is 500 nm). (3) Ultrathin encapsulation silicon oxide layer (turquoise) deposition. Corresponding top-down SEM shown in **C** (scale bar is 500 nm). (4) Erbium ions (red) implantation into the structure; (e) erbium-doped NW array after removal of the sacrificial oxide and HSQ ribbon array.

3.4. Photoluminescence Properties

To assess the proposed deterministic ion integration scheme and the potential of NW arrays for controlling the emission properties of Er^{3+} ions, room-temperature steady-state photoluminescence (PL) measurements were carried out on Er-doped SiC:O NW array structures. As shown in Figure 8, we observed a strong Er-induced PL emission around 1540 nm, which is the telecommunication C-band wavelength used in optical fibers, from Er-doped SiC:O NW, with no detectable Er^{3+} PL from the region outside the NWs (white circled points). Er-induced PL spectra ~1540 nm corresponds to the intra-4f transition ($^4I_{13/2} \rightarrow {}^4I_{15/2}$) of Er^{3+} ions, which are effectively shielded by the outer 5s and 5p electrons, resulting photostable spectra independent of annealing temperature or ambient [47,48]. Furthermore, an appreciable enhancement of the Er-induced PL was observed in the NW array structure compared to its thin-film counterpart. This enhancement can be attributed to an increase in the emission-extraction efficiency in the SiC photonic crystal structure (created by the periodic arrays of NWs), resulting from the photonic bandgap effect [46,49]. Additionally, in the case of well-passivated ultrathin NWs a reduced bulk defect-density is expected within the ion's recombination volume contributing to the observed enhanced PL [26,30]. Further details pertaining to the Er^{3+} PL behavior in SiC NW-based photonic structures are reported elsewhere [49].

Nanomaterials **2018**, *8*, 906

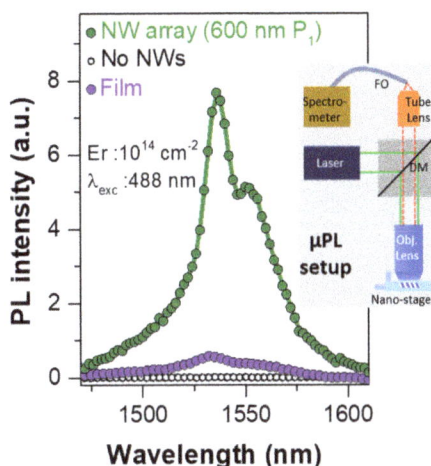

Figure 8. Room temperature Er-induced PL spectra. Er^{3+} PL ~1540 nm, corresponding to the intra-$4f$ transition ($^4I_{13/2} \rightarrow {}^4I_{15/2}$) of Er^{3+} ions, from 20 nm SiC:O NW arrays with P_1:P_2-600:150 nm, with no detectable Er^{3+} PL from the region outside the NWs (white circles). For comparison the Er^{3+} PL from a thin-film control is also shown with same Er dose (Er: 10^{14} cm^{-2} dose, 488 nm excitation). The Er^{3+} PL intensity of the thin-film was normalized to the effective area of the NW array [42,43]. Inset: A simplified schematic diagram of the home-built µPL setup composed of an argon laser, objective lens, dichroic mirror (DM), tube lens, fiber optic (FO) coupled to a spectrometer, and an InGaAs detector (see Materials and Methods).

4. Conclusions

In conclusion, herein we report a novel nanofabrication for synthesizing ultrathin self-aligned SiC or SiC:O NW arrays with on-demand positioning and tailored properties. This nanofabrication can enable the synthesis of NW arrays of a wide variety of materials, thus facilitating the study of nanostructured materials, which are difficult to produce by typical methodologies. Most importantly, this synthesis route allows for ultrathin NWs to be fabricated without the use of a lithographic-pattern-transfer technique. Additionally, we describe a fabrication scheme to deterministically integrate erbium ions into ultrathin SiC NW arrays. The high room-temperature Er^{3+} telecom-wavelength PL intensity observed from such NW arrays reveal the integration benefits of our novel nanofabrication scheme. This approach can facilitate the development of new scalable SiC NW-based systems, which can be modified towards NW-based sensing, single-photon emission, and quantum photonics applications.

Author Contributions: Conceptualization, S.G.; Methodology, V.N., M.K., B.F., N.T., S.G.; Investigation, N.T. (Nanofabrication, FTIR, AFM, XPS, PL), V.N. (Nanofabrication, PL), M.K. (Nanofabrication), B.F. (CVD, Ellipsometry), V.K. (HRTEM), S.D. (XRD), E.C. (HRTEM); Draft Preparation, B.F., S.G; Writing—Review and Editing, N.T., V.K., S.D., S.G.; Supervision, S.G.

Funding: This work was supported by the National Science Foundation through grant No ECCS-1842350. This work was also supported by the Colleges of Nanoscale Science and Engineering of SUNY Polytechnic Institute and The Research Foundation for the State University of New York. Their support is gratefully acknowledged.

Acknowledgments: The XRD data presented in this work was supported by the NSF and NIH/NIGMS via NSF award DMR-1332208. The HRSTEM data presented in this work was performed on an aberration-corrected (AC-STEM) Titan 80-300 at the Metrology department of the Colleges of Nanoscale Science and Engineering of SUNY Polytechnic Institute. Authors gratefully acknowledge engineer Corbet Johnson for the TEM sample preparation.

Conflicts of Interest: The authors declare no conflict of interest.

References

1. Cui, Y.; Wei, Q.; Park, H.; Lieber, C.M. Nanowire nanosensors for highly sensitive and selective detection of biological and chemical species. *Science* **2001**, *293*, 1289–1292. [CrossRef] [PubMed]
2. Pan, C.; Dong, L.; Zhu, G.; Niu, S.; Yu, R.; Yang, Q.; Liu, Y.; Wang, Z.L. High-resolution electroluminescent imaging of pressure distribution using a piezoelectric nanowire LED array. *Nat. Photonics* **2013**, *7*, 752–758. [CrossRef]
3. Patolsky, F.; Zheng, G.; Lieber, C.M. Nanowire-based biosensors. *Anal. Chem.* **2006**, *78*, 4260–4269. [CrossRef] [PubMed]
4. Mu, L.; Chang, Y.; Sawtelle, S.D.; Wipf, M.; Duan, X.; Reed, M.A. Silicon nanowire field-effect transistors—A versatile class of potentiometric nanobiosensors. *IEEE Access* **2015**, *3*, 287–302. [CrossRef]
5. Gao, A.; Lu, N.; Dai, P.; Li, T.; Pei, H.; Gao, X. Silicon-nanowire-based CMOS-compatible field-effect transistor nanosensors for ultrasensitive electrical detection of nucleic acids. *Nano Lett.* **2011**, *11*, 3974–3978. [CrossRef] [PubMed]
6. Patolsky, F.; Zheng, G.; Lieber, C.M. Nanowire sensors for medicine and the life sciences. *Nanomedicine (London, England)* **2006**, *1*, 51–65. [CrossRef] [PubMed]
7. Wu, H.; Chan, G.; Choi, J.W.; Ryu, I.; Yao, Y.; Mcdowell, M.T.; Lee, S.W.; Jackson, A.; Yang, Y.; Hu, L.; et al. Stable cycling of double-walled silicon nanotube battery anodes through solid–electrolyte interphase control. *Nat. Nanotechnol.* **2012**, *7*, 310–315. [CrossRef] [PubMed]
8. Gao, X.P.A.; Zheng, G.; Lieber, C.M. Subthreshold regime has the optimal sensitivity for nanowire FET biosensors. *Nano Lett.* **2010**, *10*, 547–552. [CrossRef] [PubMed]
9. Babinec, T.M.; Hausmann, B.J.M.; Khan, M.; Zhang, Y.A.; Maze, J.R.; Hemmer, P.R.; Loncar, M.A. A diamond nanowire single-photon source. *Nat. Nanotechnol.* **2010**, *5*, 195–199. [CrossRef] [PubMed]
10. Lohrmann, A.; Johnson, B.C.; McCallum, J.C.; Castelletto, S. A review on single photon sources in silicon carbide. *Rep. Prog. Phys.* **2017**, *80*, 34502. [CrossRef] [PubMed]
11. Schmidt, V.; Gosele, U. Materials science: How nanowires grow. *Science* **2007**, *316*, 698–699. [CrossRef] [PubMed]
12. Martin, C.R. Template synthesis of electronically conductive polymer nanostructures. *Acc. Chem. Res.* **1995**, *28*, 61–68. [CrossRef]
13. Xia, Y.; Yang, P.; Sun, Y.; Wu, Y.; Mayers, B.; Gates, B.; Yin, Y.; Kim, F.; Yan, H. One-dimensional nanostructures: Synthesis, characterization, and applications. *Adv. Mater.* **2003**, *15*, 353–389. [CrossRef]
14. Kwiat, M.; Cohen, S.; Pevzner, A.; Patolsky, F. Large-scale ordered 1D-nanomaterials arrays: Assembly or not? *Nano Today* **2013**, *8*, 677–694. [CrossRef]
15. Azevedo, R.G.; Jones, D.G.; Jog, A.V.; Jamshidi, B.; Myers, D.R.; Chen, L.; Fu, X.A.; Mehregany, M.; Wijesundara, M.B.J.; Pisano, A.P. A SiC MEMS resonant strain sensor for harsh environment applications. *IEEE Sens. J.* **2007**, *7*, 568–576. [CrossRef]
16. Myers, D.R. Silicon carbide resonant tuning fork for microsensing applications in high-temperature and high G-shock environments. *J. Micro/Nanolithogr. MEMS MOEMS* **2009**, *8*, 21116. [CrossRef]
17. Neudeck, P.G.; Spry, D.J.; Chen, L.Y.; Beheim, G.M.; Okojie, R.S.; Chang, C.W.; Meredith, R.D.; Ferrier, T.L.; Evans, L.J.; Krasowski, M.J.; et al. Stable electrical operation of 6H-SiC JFETs and ICs for thousands of hours at 500 °C. *IEEE Electron Device Lett.* **2008**, *29*, 456–459. [CrossRef]
18. Chen, J.; Zhang, J.; Wang, M.; Li, Y. High-temperature hydrogen sensor based on platinum nanoparticle-decorated SiC nanowire device. *Sens. Actuators B Chem.* **2014**, *201*, 402–406. [CrossRef]
19. Oliveros, A.; Guiseppi-Elie, A.; Saddow, S.E. Silicon carbide: A versatile material for biosensor applications. *Biomed. Microdevices* **2013**, *15*, 353–368. [CrossRef] [PubMed]
20. Deeken, C.R.; Esebua, M.; Bachman, S.L.; Ramshaw, B.J.; Grant, S.A. Assessment of the biocompatibility of two novel, bionanocomposite scaffolds in a rodent model. *J. Biomed. Mater. Res. Part B Appl. Biomater.* **2011**, *96*, 351–359. [CrossRef] [PubMed]
21. Frewin, C.L.; Locke, C.; Saddow, S.E.; Weeber, E.J. Single-crystal cubic silicon carbide: An in vivo biocompatible semiconductor for brain machine interface devices. In Proceedings of the 2011 Annual International Conference of the IEEE Engineering in Medicine and Biology Society, Boston, MA, USA, 30 August–3 September 2011; pp. 2957–2960. [CrossRef]

22. Fradetal, L.; Bano, E.; Attolini, G.; Rossi, F.; Stambouli, V. A silicon carbide nanowire field effect transistor for DNA detection. *Nanotechnology* **2016**, *27*, 235501. [CrossRef] [PubMed]

23. WebElements Periodic Table. Available online: https://www.webelements.com/silicon/isotopes.html (accessed on 13 August 2018).

24. Böttger, T.; Thiel, C.W.; Sun, Y.; Cone, R.L. Optical decoherence and spectral diffusion at 1.5 μm in Er^{3+}: Y_2SiO_5 versus magnetic field, temperature, and Er^{3+} concentration. *Phys. Rev. B* **2006**, *73*, 075101. [CrossRef]

25. Cheng, G.; Chang, T.H.; Qin, Q.; Huang, H.; Zhu, Y. Mechanical properties of silicon carbide nanowires: Effect of size-dependent defect density. *Nano Lett.* **2014**, *14*, 754–758. [CrossRef] [PubMed]

26. Estes, M.J.; Moddel, G. Luminescence from amorphous silicon nanostructures. *Phys. Rev. B* **1996**, *54*, 14633–14642. [CrossRef]

27. Orchard, J.R.; Woodhead, C.; Wu, J.; Tang, M.; Beanland, R.; Noori, Y.; Liu, H.; Young, R.J.; Mowbray, D.J. Silicon-based single quantum dot emission in the telecoms C-band. *ACS Photonics* **2017**, *4*, 1740–1746. [CrossRef]

28. Hanson, R. Quantum information: Mother nature outgrown. *Nat. Mater.* **2009**, *8*, 368. [CrossRef] [PubMed]

29. Ford, B.; Tabassum, N.; Nikas, V.; Gallis, S. Strong photoluminescence enhancement of silicon oxycarbide through defect engineering. *Materials* **2017**, *10*, 446. [CrossRef] [PubMed]

30. Tabassum, N.; Nikas, V.; Ford, B.; Huang, M.; Kaloyeros, A.E.; Gallis, S. Time-resolved analysis of the White photoluminescence from chemically synthesized SiC_xO_y thin films and nanowires. *Appl. Phys. Lett.* **2016**, *109*, 43104. [CrossRef]

31. Nikas, V.; Tabassum, N.; Ford, B.; Smith, L.; Kaloyeros, A.E.; Gallis, S. Strong visible light emission from silicon-oxycarbide nanowire arrays prepared by electron beam lithography and reactive ion etching. *J. Mater. Res.* **2015**, *30*, 3692–3699. [CrossRef]

32. Pawbake, A.; Mayabadi, A.; Waykar, R.; Kulkarni, R.; Jadhavar, A.; Waman, V.; Parmar, J.; Bhattacharyya, S.; Ma, Y.R.; Devan, R.; et al. Growth of boron doped hydrogenated nanocrystalline cubic silicon carbide (3C-SiC) films by Hot Wire-CVD. *Mater. Res. Bull.* **2016**, *76*, 205–215. [CrossRef]

33. Zekentes, K.; Rogdakis, K. SiC nanowires: Material and devices. *J. Phys. D Appl. Phys.* **2011**, *44*, 133001. [CrossRef]

34. Gallis, S.; Nikas, V.; Huang, M.; Eisenbraun, E.; Kaloyeros, A.E. Comparative study of the effects of thermal treatment on the optical properties of hydrogenated amorphous silicon-oxycarbide. *J. Appl. Phys.* **2007**, *102*, 024302. [CrossRef]

35. Gallis, S.; Nikas, V.; Eisenbraun, E.; Huang, M.; Kaloyeros, A.E. On the effects of thermal treatment on the composition, structure, morphology, and optical properties of hydrogenated amorphous silicon-oxycarbide. *J. Mater. Res.* **2009**, *24*, 2561–2573. [CrossRef]

36. Calcagno, L.; Musumeci, P.; Roccaforte, F.; Bongiorno, C.; Foti, G. Crystallisation mechanism of amorphous silicon carbide. *Appl. Surf. Sci.* **2001**, *184*, 123–127. [CrossRef]

37. Madapura, S. Heteroepitaxial growth of SiC on Si(100) and (111) by chemical vapor deposition using trimethylsilane. *J. Electrochem. Soc.* **1999**, *146*, 1197–1202. [CrossRef]

38. Steckl, A.J.; Devrajan, J.; Tlali, S.; Jackson, H.E.; Tran, C.; Gorin, S.N.; Ivanova, L.M. Characterization of 3C–SiC crystals grown by thermal decomposition of methyltrichlorosilane. *Appl. Phys. Lett.* **1996**, *69*, 3824–3826. [CrossRef]

39. Shaffer, P.T.B.; Naum, R.G. Refractive index and dispersion of beta silicon carbide. *J. Opt. Soc. Am.* **1969**, *59*, 1498. [CrossRef]

40. Persson, C.; Lindefelt, U. Detailed band structure for 3C-, 2H-, 4H-, 6H-SiC, and Si around the fundamental band gap. *Phys. Rev. B Condens. Matter Mater. Phys.* **1996**, *54*, 10257–10260. [CrossRef]

41. Dey, S.; Tapily, K.; Consiglio, S.; Clark, R.D.; Wajda, C.S.; Leusink, G.L.; Woll, A.R.; Diebold, A.C. Role of Ge and Si substrates in higher-k tetragonal phase formation and interfacial properties in cyclical atomic layer deposition-anneal $Hf_{1-x}Zr_xO_2/Al_2O_3$ thin film stacks. *J. Appl. Phys.* **2016**, *120*. [CrossRef]

42. Iwanowski, R.; Fronc, K.; Paszkowicz, W.; Heinonen, M. XPS and XRD study of crystalline 3C-SiC grown by sublimation method. *J. Alloys Compd.* **1999**, *286*, 143–147. [CrossRef]

43. Hall, W.H. X-ray line broadening in metals. *Proc. Phys. Soc. Sect. A* **1949**, *62*, 741–743. [CrossRef]

44. Wu, R.; Zhou, K.; Yue, C.Y.; Wei, J.; Pan, Y. Recent progress in synthesis, properties and potential applications of SiC nanomaterials. *Prog. Mater. Sci.* **2015**, *72*, 1–60. [CrossRef]

45. Gao, L.; Zhong, H.; Chen, Q. Synthesis of 3C–SiC nanowires by reaction of poly (ethylene terephthalate) waste with SiO_2 microspheres. *J. Alloys Compd.* **2013**, *566*, 212–216. [CrossRef]

46. Noda, S.; Fujita, M.; Asano, T. Spontaneous-emission control by photonic crystals and nanocavities. *Nat. Photonics* **2007**, *1*, 449–458. [CrossRef]

47. Polman, A. Erbium implanted thin film photonic materials. *J. Appl. Phys.* **1997**, *82*, 1–39. [CrossRef]

48. Nikas, V.; Gallis, S.; Huang, M.; Kaloyeros, A.E. Thermal annealing effects on photoluminescence properties of carbon-doped silicon-rich oxide thin films implanted with erbium. *J. Appl. Phys.* **2011**, *109*, 093521. [CrossRef]

49. Tabassum, N.; Nikas, V.; Ford, B.; Crawford, E.; Gallis, S. Engineering Er^{3+} placement and emission through chemically-synthesized self-aligned SiC: Ox nanowire photonic crystal structures. *arXiv*, 2017; arXiv:1707.05738.

nanomaterials

MDPI

Article

Bendable Single Crystal Silicon Nanomembrane Thin Film Transistors with Improved Low-Temperature Processed Metal/n-Si Ohmic Contact by Inserting TiO$_2$ Interlayer

Jiaqi Zhang †, Yi Zhang †, Dazheng Chen, Weidong Zhu, He Xi, Jincheng Zhang, Chunfu Zhang * and Yue Hao

State Key Discipline Laboratory of Wide Band Gap Semiconductor Technology, School of Microelectronics, Xidian University, 2 South Taibai Road, Xi'an 710071, China; 18031362893@163.com (J.Z.); hyper_sys@163.com (Y.Z.); dzchen@xidian.edu.cn (D.C.), wdzhu@xidian.edu.cn (W.Z.); hxi@xidian.edu.cn (H.X.); jchzhang@xidian.edu.cn (J.Z.); yhao@xidian.edu.cn (Y.H.)
* Correspondence: cfzhang@xidian.edu.cn; Tel.: +86-88201759-818
† These authors contributed equally to this work.

Received: 21 November 2018; Accepted: 13 December 2018; Published: 16 December 2018

Abstract: Bendable single crystal silicon nanomembrane thin film transistors (SiNMs TFTs), employing a simple method which can improve the metal/n-Silicon (Si) contact characteristics by inserting the titanium dioxide (TiO$_2$) interlayer deposited by atomic layer deposition (ALD) at a low temperature (90 °C), are fabricated on ITO/PET flexible substrates. Current-voltage characteristics of titanium (Ti)/insertion layer (IL)/n-Si structures demonstrates that they are typically ohmic contacts. X-ray photoelectron spectroscopy (XPS) results determines that TiO$_2$ is oxygen-vacancies rich, which may dope TiO$_2$ and contribute to a lower resistance. By inserting TiO$_2$ between Ti and n-Si, I$_{ds}$ of bendable single crystal SiNMs TFTs increases 3–10 times than those without the TiO$_2$ insertion layer. The fabricated bendable devices show superior flexible properties. The TFTs, whose electrical properties keeps almost unchanged in 800 cycles bending with a bending radius of 0.75 cm, obtains the durability in bending test. All of the results confirm that it is a promising method to insert the TiO$_2$ interlayer for improving the Metal/n-Si ohmic contact in fabrication of bendable single crystal SiNMs TFTs.

Keywords: thin film transistor; single-crystal Si nanomembrane (Si NMs); TiO$_2$ insertion layer; ohmic contact

1. Introduction

Flexible electronics is an important development direction in the field of future electronics. Scientists can use flexible materials to fabricate advanced electronic devices, such as transistor arrays for optional folding and stretching, bendable flexible screens, or some sensors which can be integrated on the clothing [1–9]. Advances in various flexible electronic technologies, including solar cells, sensors and displays, have been driven by the use of flexible organic materials. However, organic-based semiconductors suffer from poor device performance due to their low carrier mobility and their chemical/thermal instability. Recently, the discovery of single-crystal Si nanomembranes (SiNMs) has fascinated the flexible electronics community because of their high carrier mobility, stable chemical/thermal properties and flexibility. Particularly, SiNMs released from silicon-on-insulator (SOI) become one of the best choices owing to their outstanding electrical properties, mature fabrication techniques and commercial feasibility at relatively lower cost [10–14].

For SiNMs, effectively doping them to form effective ohmic contacts and realizing low contact resistivity are both critical in realizing high speed operation. Nevertheless, many flexible substrates are soft and have very low processing temperature tolerance. For example, ITO/PET substrates just can withstand the highest temperature as low as 150 °C [11]. Hence, the traditional high-temperature processing cannot be directly used. This challenge has been partially overcome by employing a pre-doped (ion implantation and annealing before SiNMs release) SiNMs transfer and gate-last TFTs fabrication process [10,11]. However, a low-temperature process to achieve a lower Metal/Si Ohmic contact is still urgently pursued.

On the basis of the works above, we reports a simple method that using ALD technology deposits TiO_2 at a low temperature of 90 °C to further improve the contact between the source/drain regions and metal electrodes. Current-voltage characteristics of Ti/insertion layer (IL)/n-Si structures, XPS results of TiO_2 and the normalized current-voltage characteristics of bendable single crystal silicon TFTs with different cycles of TiO_2 are obtained and described in detail. By inserting titanium dioxide, good ohmic contacts are formed between the source/drain regions and metal electrodes. I_{ds} of bendable single crystal SiNMs TFTs increases 3–10 times than those without the TiO_2 insertion layer. The TFTs, whose electrical properties keeps almost unchanged in 800 cycles bending with a bending radius of 0.75 cm, obtains the durability in bending test.

2. Materials and Methods

2.1. Device Fabrication

Figure 1 schematically illustrates the cross section of the two kinds of bendable single crystal silicon TFTs built on ITO/PET substrates and the devices fabrication process. Figure 1a shows structure schematic of a bendable single crystal silicon TFT without inserting TiO_2 and Figure 1b shows structure schematic of a bendable single crystal silicon TFT with inserting different cycles of TiO_2. Figure 1c shows that the devices fabrication process was started with silicon-on-insulator wafer (SOI) (Soitec by Smartcut with 200 nm top Si which is doped boron whose level is 1×10^{14} cm^{-3} and 200 nm buried oxide). In our process of making n-channel TFTs, the source/drain regions were first formed on the SOI substrate via phosphorus ion implantation with a dose of 5×10^{15} cm^{-2} and an energy of 30 keV followed by annealing in RTP at 1000 °C for 20 s in N_2 ambient to activate the implanted dopants. The doping concentration of the contact area is ~10^{19} cm^{-3}. SOI wafer was patterned by lithography to form the hole patterns. RIE was used to etch the holes. 33% hydrofluoric acid (HF) etched buried oxide (SiO_2) through a lot of etching holes above SiO_2 [15–19]. Completely etching buried oxide took about 48 h so that the top Si dropped on the bottom silicon substrate by Van der Waals force [20–22]. Deionized water was used to wash away residual HF to be ready for transfer. In order to avoid being dislocated and scattered of the SiNMs which was immersed in HF, we designed the mask including a lot of 1×1 cm units so that the SiNMs formed a very large area. Even if the buried oxygen was completely etched off, there was no displacement and scatter of the SiNMs on the bottom Si. At the same time, this design increased the transfer area reaching 1 cm^2. A flat piece of polydimethylsiloxane (PDMS) was brought into conformal contact with the top surface of the wafer and then rapidly peeled back to pick up SiNMs. The interaction between SiNMs and PDMS is sufficient to pick up SiNMs with good efficiency, nearly 100%. A flexible ITO/PET substrate served as the target substrate. The target substrate was washed with acetone and alcohol, rinsed with deionized water and then dried with a stream of nitrogen. Treating the ITO/PET substrate with a short O_2 plasma (20 sccm O_2 flow with 50 W rf power for 10 s) promoted adhesion between it and a spin coating dielectric layer of epoxy (4000 rpm for 30 s of SU8-2002). Then the SiNMs on PDMS was brought against the epoxy layer (SU8-2002). SiNMs was transferred to the ITO/PET substrate by gently pressing and slowly peeling up PDMS. The epoxy layer was cured at 100 °C for 1 min, exposed to UV light from the backside of the sample for 10 s and finally post-baked at 100 °C for 1 min. Photolithography defined a pattern on the substrate. RIE etched the Si to form MESA. O_2 plasma was used to remove photoresist. After 5% HF solution immersion for 2 min and de-ionized water rinse for 2 min, N_2 gun blew dry the samples. Then the samples were immediately loaded into PicosunTM R-200 Advanced ALD chamber. Titanium tetrakis (dimethylamide) (TDMATi) was used as Ti source and H_2O was used as oxygen source. During the process, TDMATi source bottle was heated to 120 °C and the N_2 carrier flow was set to 15 standard cubic centimeter per minute (sccm), pulse time and purge time were 100 ms and 40 s respectively. For oxygen source, the N_2 carrier flow was set to 15 sccm, pulse time was 100 ms and purge time was 40 s. The whole deposition process was carried out at 90 °C. There were two different cycles of 5 and 10. Using these two conditions formed ~0.5 and ~1 nm TiO_2 on the SiNMs, respectively. Photolithography defined a pattern on the samples. Put them into a PRO LINE PVD 75 SYSTEM (Kurt J. Lesker Company, Pittsburgh, PA, USA) to deposit Ti (100 nm) by electron beam evaporation. Finally, the source and drain metal contacts formed on the low-resistive source and drain regions followed by lift-off without any further thermal treatment.

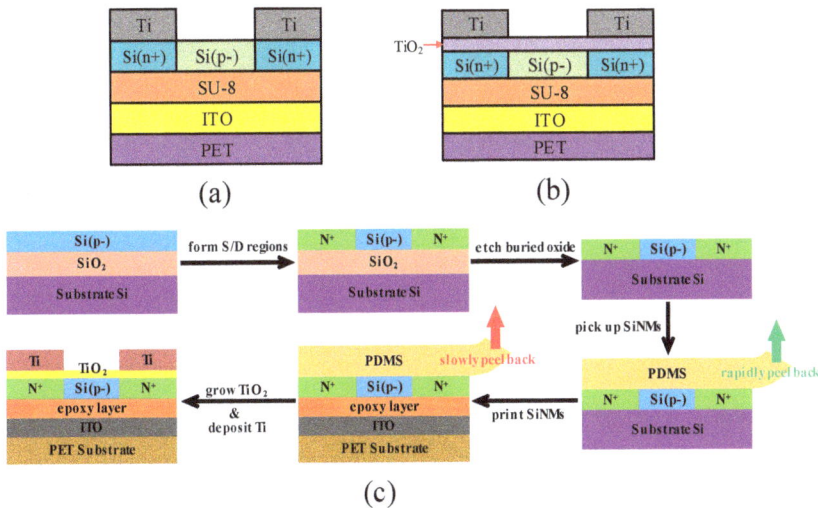

Figure 1. (**a**) Structure schematic of a bendable single crystal silicon TFT without inserting TiO_2. (**b**) Structure schematic of a bendable single crystal silicon TFT with inserting different cycles of TiO_2. (**c**) Schematic illustration of fabrication process for bendable single crystal silicon TFT.

2.2. Device Characterization

Etching depth of SiNMs was measured by Stylus Profiler (Bruker Dektak XT, Bremen, Germany). The XPS testing of TiO_2 samples was performed on Thermo escalab 250Xi (Thermo Fisher Scientific, Waltham, MA, USA) using monochromatic Al-Ka (1486.6 eV) as the radiation source. The I–V characteristics of Ti/insertion layer (IL)/n-Si metal-insulator-semiconductor (MIS) structures and bendable single crystal silicon TFTs were both measured by using Keithley 1500 semiconductor characterization system (Tektronix, Inc., Beaverton, OR, USA). All the measurements were performed under ambient atmosphere at room temperature without encapsulation.

3. Results and Discussion

Figure 2a presents a schematic cross-sectional view of the Ti/insertion layer (IL)/n-Si metal-insulator-semiconductor (MIS) structures with different ALD cycles. Figure 2b shows a high magnification optical images of the MIS structure whose Ti pad diameter is 300 μm. Figure 2c shows the I–V characteristics of Ti/insertion layer (IL)/n-Si (doping level is ~10^{19} cm^{-3}) MIS structure with different ALD cycles. It is obvious that the characteristic curve of devices without TiO_2 (0 cycle) is curving which is a typical Schottky contact. It indicates that even if the source/drain area is heavily doped, Ti/n-Si (doping level is ~10^{19} cm^{-3}) will not form a good ohmic contact without any treatment. Amazingly, the characteristic curves of devices which are deposited by the TiO$_2$ insertion layer are all typically good ohmic contacts. The contact resistance of 5 cycle is the smallest and the contact resistance of 10 cycle is the second smallest. However, with the increase of the cycle (the thickness of TiO$_2$), the contact resistance becomes greater and greater. The explanation for this is that the thickness of titanium dioxide exceeds a certain value leading to carriers passing through the insertion layer at lower tunneling rate [23,24]. This will reduce the current, with presenting a larger contact resistance. Therefore, we chose 5 and 10 cycles of titanium dioxide inserting into the TFTs to compare those without titanium dioxide insertion layer.

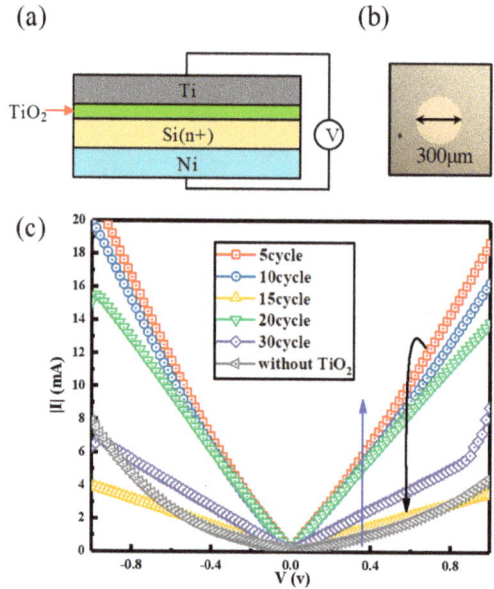

Figure 2. (a) Schematic cross-sectional view of the Ti/insertion layer (IL)/n-Si metal-insulator-semiconductor (MIS) structures with different ALD cycles. (b) A high magnification optical images of the MIS structure whose Ti pad diameter is 300 μm. (c) The I–V characteristics of Ti/insertion layer (IL)/n-Si (doping level is ~10^{19} cm^{-3}) MIS structure with different ALD cycles.

Figure 3a shows the XPS results of Ti 2p of samples for 20 cycles and 300 cycles of ALD process and all the XPS results are calibrated with C 1 s peak at 284.8 eV [23,24]. The two curves have two distinct peaks at about 458–459 eV and 464–465 eV, which represent for Ti 2p3/2 and Ti 2p1/2 peaks and are consistent with typical values of TiO$_2$. Figure 3b shows that O 1 s results with thin TiO$_2$, shoulder left to O 1 s peak is obvious. This shoulder located at 531.5 eV represents for oxygen vacancies and it indicates that thin sample has more oxygen vacancies. The illustration in the upper left corner shows the fitting curve of 300 cycles. It can be seen that the peak of oxygen vacancies is not obvious. In conclusion, there are some oxygen vacancies in thin TiO$_2$ film, which will dope TiO$_2$ and make it more conductive [24]. Due to the presence of a donor band related to oxygen vacancies which can provide more electrons, TiO$_2$ behaves as an n-type semiconductor and exhibits good electrical conductivity.

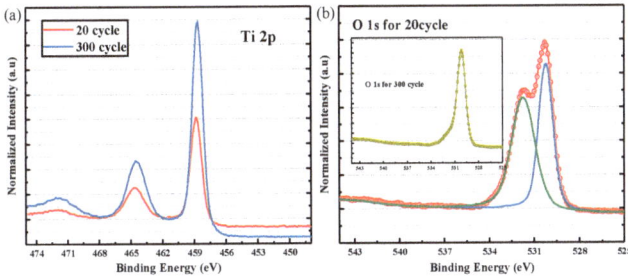

Figure 3. (**a**) The XPS results of Ti 2p of samples for 20 cycles and 300 cycles of ALD process. (**b**) The XPS results of O 1 s with thin TiO_2, shoulder left to O 1 s peak is obvious. The illustration in the upper left corner shows the fitting curve of 300 cycles.

Figure 4a–d presents the normalized current-voltage characteristics of TFTs with different cycles of TiO_2. As shown in the figure, the I_{ds} of 0 cycle is the smallest. The I_{ds} of 5 cycle is the greatest. Figure 4d shows the I_{ds} of 0, 5, 10 cycle at $V_{gs} = 5$ v. It presents that the current of 10 cycles is three times the 0 cycle one. The current of 5 cycles is ten times the 0 cycle one. These results are consistent with results of Figure 2c. The reason is that inserting the titanium dioxide of appropriate thickness between Ti and n-Si can effectively restrain the Fermi energy level pinning effect of n-type silicon so as to improve the Ti/n-Si ohmic contact, reduce the contact resistance and increase the current driving ability.

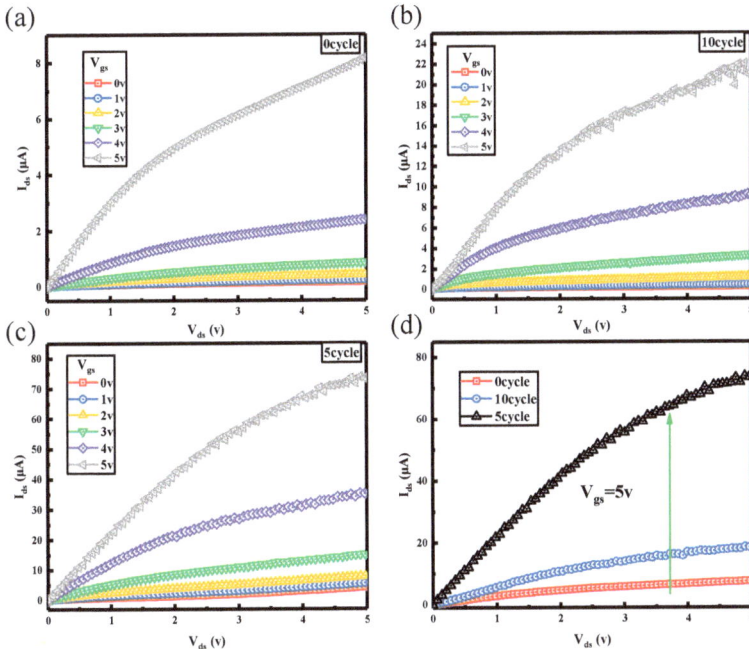

Figure 4. The normalized current-voltage characteristics of the thin film transistor devices on PET substrate with different cycle of TiO_2. (**a**) The thin film transistor devices on PET substrate with 0 cycle of TiO_2. (**b**) The thin film transistor devices on PET substrate with 10 cycles of TiO_2. (**c**) The thin film transistor devices on PET substrate with 5 cycles of TiO_2. (**d**) Shows the normalized current-voltage characteristics of the thin film transistor devices on PET substrate whose I_{ds} of 0, 5, 10 cycles at $V_{gs} = 5$ v.

Figure 5a presents transfer characteristics of TFTs with different cycles of TiO_2. $I_{on/off}$ of 0 cycle is 10^2, 5 cycles is 10^4 and 10 cycles is 10^3. I_{on} of 0 cycle is smallest, 5 cycles is largest and 10 cycles is medium. This result is consistent with results of Figures 2c and 4d. Thin TiO_2 which restrains the Fermi energy level pinning effect of n-Si can improve the interface contact between Ti/n-Si contributing to increase I_{ds} [23–25]. I_{off} of 0 cycle is largest, 5 cycles is smallest and 10 cycles is medium. For transistors, the smaller the I_{off}, the better the performance of the device. After analysis, it may be that ALD growth of titanium oxide plays anneal role so that the interface contact between epoxy layer (SU8-2002) and SiNMs is better than that without depositing TiO_2. It promotes the ability to control current of gate electrode. To test this conjecture, we put the TFT without depositing TiO_2 into ALD annealing at 90 °C for 5 min (the time of depositing 5 cycles TiO_2) and then compared the transfer characteristics with unannealed one. Results are as Figure 5b shown, the I_{off} of without TiO_2 but with annealing is an order of magnitude smaller than that without TiO_2 and without annealing, which is consistent with results of Figure 5a. Hence, it indicates that even if annealing at 90 °C can also improve the interface contact between epoxy layer (SU8-2002) and SiNMs with reducing interface defects and enhancing the ability to control current of gate electrode to reduce I_{off}. Thus, the combination of the inserted TiO_2 layer and very low-temperature annealing process improves the metal/Si ohmic contact.

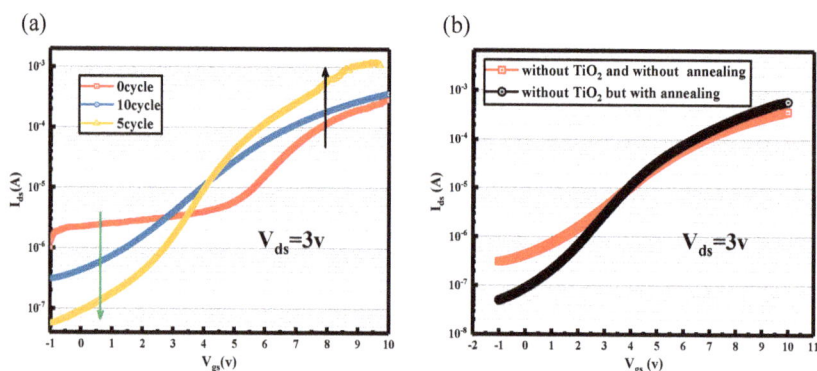

Figure 5. (a) Transfer characteristics of TFTs on ITO/PET substrates with different cycles of TiO_2. (b) Transfer characteristics of the annealed TFT which are not deposited TiO_2 on ITO/PET substrates compared with one without annealing.

Figure 6a is an image of the flexible TFTs fixed on a probe station with the bending radius of 0.75 cm. Figure 6b shows the digital photographs of the flexible TFTs. Figure 6c shows an image of one device of TFTs under an optical microscope. Figure 6d shows a bending test of the flexible TFTs with a bending radius of 0.75 cm. Obtaining the durability in bending conditions is essential to wearable applications [26]. The electrical properties of the devices do not change significantly in 800 cycles with a bending radius of 0.75 cm, confirming a reliable performance for flexible operations.

Nanomaterials **2018**, *8*, 1060

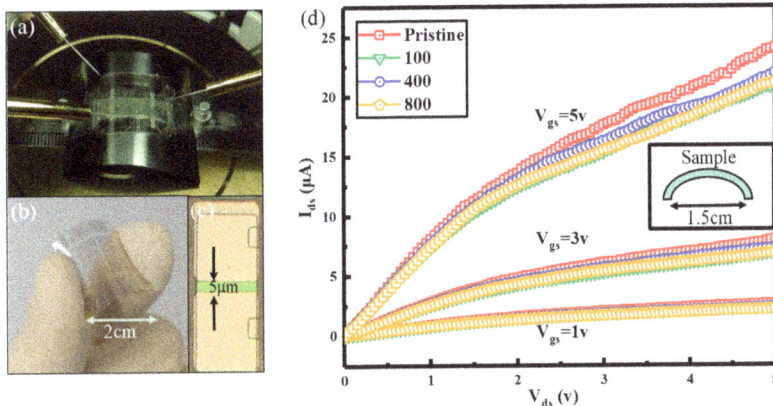

Figure 6. (**a**) The image of the flexible TFTs fixed on a probe station with the bending radius of 0.75 cm. (**b**) Shows the digital photographs of the flexible TFTs. (**c**) Shows an image of one device of TFTs under an optical microscope. (**d**) Shows the electrical characteristics of the flexible TFTs with a bending radius of 0.75 cm after being bent several times.

4. Conclusions

A simple method was used to improve Ti/n-Si contact characteristics by inserting ALD deposited TiO_2. The fabrication temperature of inserting ALD deposited TiO_2 is below 120 °C which is the highest temperature plastic can withstand. Hence, we were able to combine this method into the manufacture of flexible thin film transistors. The results show that this method is not only feasible but also effective. It can change the Schottky contact into a satisfactory ohmic contact so that increase current drive capability. The electrical properties of the flexible TFTs do not change significantly with bending 800 cycles, confirming a reliable performance for flexible operations.

Author Contributions: C.Z. conceived the idea, designed the experiment and guided the experiment. J.Z. conducted most of the device fabrication and data collection and wrote the manuscript; C.Z. and Y.Z. revised the manuscript; D.C., W.Z., H.X. helped the device measurement, J.Z. helped the data analysis. Y.H. supervised the group. All authors read and approved the manuscript.

Funding: We thank the Fundamental Research Funds for the National 111 Center (Grant No. B12026), Natural Science Foundation of Shaanxi Province (Grant No. 2017JM6049), the Central Universities (Grant no. JBX171103), Natural Science Foundation of China (61604119) and Class General Financial Grant from the China Postdoctoral Science Foundation (Grant No. 2016M602771).

Conflicts of Interest: The authors declare no conflicts of interest.

References

1. Kim, J.; Lee, M.; Shim, H.J.; Ghaffari, R.; Cho, H.R.; Son, D.; Jung, Y.H.; Soh, M.; Choi, C.; Jung, S.; et al. Stretchable silicon nanoribbon electronics for skin prosthesis. *Nat. Commun.* **2014**, *5*, 1–11. [CrossRef] [PubMed]

2. Ying, M.; Bonifas, A.P.; Lu, N.; Su, Y.; Li, R.; Cheng, H.; Ameen, A.; Huang, Y.; Rogers, J.A. Silicon nanomembranes for fingertip electronics. *Nanotechnology* **2012**, *23*, 344004. [CrossRef] [PubMed]

3. Park, M.; Kim, M.S.; Park, Y.K.; Ahn, J.H. Si membrane based tactile sensor with active matrix circuitry for artificial skin applications. *Appl. Phys. Lett.* **2015**, *106*, 043502. [CrossRef]

4. Won, S.M.; Kim, H.S.; Lu, N.; Kim, D.G.; Del Solar, C.; Duenas, T.; Ameen, A.; Rogers, J.A. Piezoresistive strain sensors and multiplexed arrays using assemblies of single-crystalline silicon nanoribbons on plastic substrates. *IEEE Trans. Electron Devices* **2011**, *58*, 4074–4078. [CrossRef]

5. Ahn, J.H.; Kim, H.S.; Lee, K.J.; Jeon, S.; Kang, S.J.; Sun, Y.; Nuzzo, R.G.; Rogers, J.A. Heterogeneous three-dimensional electronics by use of printed semiconductor nanomaterials. *Science* **2006**, *314*, 1754–1757. [CrossRef] [PubMed]

Nanomaterials **2018**, *8*, 1060

6. Yang, S.; Lu, N. Gauge factor and stretchability of silicon-on-polymer strain gauges. *Sensors* **2013**, *13*, 8577–8594. [CrossRef]

7. Lu, N.; Kim, D.-H. Flexible and Stretchable Electronics Paving the Way for Soft Robotics. *Soft Robot.* **2014**, *1*, 53–62. [CrossRef]

8. Kim, D.H.; Ahn, J.H.; Kim, H.S.; Lee, K.J.; Kim, T.H.; Yu, C.J.; Nuzzo, R.G.; Rogers, J.A. Complementary logic gates and ring oscillators on plastic substrates by use of printed ribbons of single-crystalline silicon. *IEEE Electron Device Lett.* **2008**, *29*, 73–76. [CrossRef]

9. Ahn, J.H.; Kim, H.S.; Menard, E.; Lee, K.J.; Zhu, Z.; Kim, D.H.; Nuzzo, R.G.; Rogers, J.A.; Amlani, I.; Kushner, V.; et al. Bendable integrated circuits on plastic substrates by use of printed ribbons of single-crystalline silicon. *Appl. Phys. Lett.* **2007**, *90*, 2005–2008. [CrossRef]

10. Sun, L.; Qin, G.; Seo, J.H.; Celler, G.K.; Zhou, W.; Ma, Z. 12-GHz thin-film transistors on transferrable silicon nanomembranes for high-performance flexible electronics. *Small* **2010**, *6*, 2553–2557. [CrossRef]

11. Yuan, H.C.; Celler, G.K.; Ma, Z. 7.8-GHz flexible thin-film transistors on a low-temperature plastic substrate. *J. Appl. Phys.* **2007**, *102*, 034501. [CrossRef]

12. Torres Sevilla, G.A.; Almuslem, A.S.; Gumus, A.; Hussain, A.M.; Cruz, M.E.; Hussain, M.M. High performance high-κ/metal gate complementary metal oxide semiconductor circuit element on flexible silicon. *Appl. Phys. Lett.* **2016**, *108*, 094102. [CrossRef]

13. Gupta, S.; Navaraj, W.T.; Lorenzelli, L.; Dahiya, R. Ultra-thin chips for high-performance flexible electronics. *npj Flex. Electron.* **2018**, *2*, 8. [CrossRef]

14. Zhang, K.; Seo, J.H.; Zhou, W.; Ma, Z. Fast flexible electronics using transferrable silicon nanomembranes. *J. Physics D Appl. Phys.* **2012**, *45*, 143001. [CrossRef]

15. Cohen, G.M.; Mooney, P.M.; Paruchuri, V.K.; Hovel, H.J. Dislocation-free strained silicon-on-silicon by in-place bonding. *Appl. Phys. Lett.* **2005**, *86*, 251902. [CrossRef]

16. Song, E.; Guo, Q.; Huang, G.; Jia, B.; Mei, Y. Bendable Photodetector on Fibers Wrapped with Flexible Ultrathin Single Crystalline Silicon Nanomembranes. *ACS Appl. Mater. Interfaces* **2017**, *9*, 12171–12175. [CrossRef]

17. Guo, Q.; Fang, Y.; Zhang, M.; Huang, G.; Chu, P.K.; Mei, Y.; Di, Z.; Wang, X. Wrinkled Single-Crystalline Germanium Nanomembranes for Stretchable Photodetectors. *IEEE Trans. Electron Devices* **2017**, *64*, 1985–1990. [CrossRef]

18. Roberts, M.M.; Klein, L.J.; Savage, D.E.; Slinker, K.A.; Friesen, M.; Celler, G.; Eriksson, M.A.; Lagally, M.G. Elastically relaxed free-standing strained-silicon nanomembranes. *Nat. Mater.* **2006**, *5*, 388–393. [CrossRef]

19. Song, E.; Fang, H.; Jin, X.; Zhao, J.; Jiang, C.; Yu, K.J.; Zhong, Y.; Xu, D.; Li, J.; Fang, G.; et al. Thin, Transferred Layers of Silicon Dioxide and Silicon Nitride as Water and Ion Barriers for Implantable Flexible Electronic Systems. *Adv. Electron. Mater.* **2017**, *3*, 1700077. [CrossRef]

20. Meitl, M.A.; Zhu, Z.T.; Kumar, V.; Lee, K.J.; Feng, X.; Huang, Y.Y.; Adesida, I.; Nuzzo, R.G.; Rogers, J.A. Transfer printing by kinetic control of adhesion to an elastomeric stamp. *Nat. Mater.* **2006**, *5*, 33–38. [CrossRef]

21. Carlson, A.; Bowen, A.M.; Huang, Y.; Nuzzo, R.G.; Rogers, J.A. Transfer printing techniques for materials assembly and micro/nanodevice fabrication. *Adv. Mater.* **2012**, *24*, 5284–5318. [CrossRef] [PubMed]

22. Menard, E.; Lee, K.J.; Khang, D.Y.; Nuzzo, R.G.; Rogers, J.A. A printable form of silicon for high performance thin film transistors on plastic substrates. *Appl. Phys. Lett.* **2004**, *84*, 5398–5400. [CrossRef]

23. Stefanov, P.; Shipochka, M.; Stefchev, P.; Raicheva, Z.; Lazarova, V.; Spassov, L. XPS characterization of TiO_2 layers deposited on quartz plates. *J. Phys. Conf. Ser.* **2008**, *100*, 012039. [CrossRef]

24. Zhang, Y.; Liu, Y.; Han, G.; Liu, H.; Hao, Y. Improving metal/n-Ge ohmic contact by inserting TiO2 deposited by PEALD. *Micro Nano Lett.* **2018**, *13*, 801–803. [CrossRef]

25. Hobbs, C.; Fonseca, L.; Dhandapani, V.; Samavedam, S.; Taylor, B.; Grant, J.; Dip, L.; Triyoso, D.; Hegde, R.; Gilmer, D.; et al. Fermi level pinning at the polySi/metal oxide interface. In Proceedings of the 2003 Symposium on VLSI Technology. Digest of Technical Papers (IEEE Cat. No.03CH37407), Kyoto, Japan, 10–12 June 2003; pp. 3–4. [CrossRef]

26. Rim, Y.S.; Yang, Y.; Bae, S.H.; Chen, H.; Li, C.; Goorsky, M.S.; Yang, Y. Ultrahigh and Broad Spectral Photodetectivity of an Organic-Inorganic Hybrid Phototransistor for Flexible Electronics. *Adv. Mater.* **2015**, *27*, 6885–6891. [CrossRef] [PubMed]

nanomaterials

MDPI

Article

Growth and Self-Assembly of Silicon–Silicon Carbide Nanoparticles into Hybrid Worm-Like Nanostructures at the Silicon Wafer Surface

Manuel Alejandro Perez-Guzman [1,*], Rebeca Ortega-Amaya [2,*], Yasuhiro Matsumoto [2], Andres Mauricio Espinoza-Rivas [3], Juan Morales-Corona [4], Jaime Santoyo-Salazar [5] and Mauricio Ortega-Lopez [2]

1   Programa de Doctorado en Nanociencias y Nanotecnología, Centro de Investigación y de Estudios Avanzados del IPN, Av. IPN 2508, Col. San Pedro Zacatenco, Ciudad de México 07360, Mexico
2   SEES, Departamento de Ingeniería Eléctrica, Centro de Investigación y de Estudios Avanzados del IPN, Av. IPN 2508, Col. San Pedro Zacatenco, Ciudad de México 07360, Mexico; ymatsumo@cinvestav.mx (Y.M.); ortegal@cinvestav.mx (M.O.-L.)
3   Departamento de Ingeniería Eléctrica, Universidad Tecnológica de México-UNITEC MÉXICO-Campus Cuitláhuac, Norte 67 2346, Col. San Salvador Xochimanca, Ciudad de México 02870, Mexico; mauricio.inu@gmail.com
4   Departamento de Física, Universidad Autónoma Metropolitana, Unidad Iztapalapa, Av. San Rafael Atlixco 186, Col. Vicentina, Ciudad de México 09340, Mexico; jmor@xanum.uam.mx
5   Departamento de Física, Centro de Investigación y de Estudios Avanzados del IPN, Av. IPN 2508, Col. San Pedro Zacatenco, Ciudad de México 07360, Mexico; jsantoyo@fis.cinvestav.mx
*   Correspondence: maperez@cinvestav.mx (M.A.P.-G.); ortegaa@cinvestav.mx (R.O.-A.); Tel.: +52-55-5747-3800 (ext. 6260) (M.A.P.-G. & R.O.-A)

Received: 16 October 2018; Accepted: 16 November 2018; Published: 20 November 2018

Abstract: This work describes the growth of silicon–silicon carbide nanoparticles (Si–SiC) and their self-assembly into worm-like 1D hybrid nanostructures at the interface of graphene oxide/silicon wafer (GO/Si) under Ar atmosphere at 1000 °C. Depending on GO film thickness, spread silicon nanoparticles apparently develop on GO layers, or GO-embedded Si–SiC nanoparticles self-assembled into some-micrometers-long worm-like nanowires. It was found that the nanoarrays show that carbon–silicon-based nanowires (CSNW) are standing on the Si wafer. It was assumed that Si nanoparticles originated from melted Si at the Si wafer surface and GO-induced nucleation. Additionally, a mechanism for the formation of CSNW is proposed.

Keywords: silicon; silicon carbide; nanoparticles; nanowires; graphene oxide; self-assembly; thermal reduction

---

1. Introduction

Nanostructures of organic and inorganic materials display singular physicochemical properties that depend on their chemical composition and morphology [1–3]. Recent advances on synthesis methods, and morphological and phase composition control [4,5] have led to the development of novel multifunctional nanomaterials for applications in medicine [6], energy conversion and storage [7], environmental remediation [8], and electronics [9].

The synthesis of the core–shell hybrid inorganic–organic nanostructures is intensively pursued because it combines both the inorganic core functionality and its stability under operation; moreover, the core itself may be designed to allow for an additional functionalization, expanding its technological applications [10,11]. In the last years, a number of graphene or graphene oxide (GO)-coated transition metal oxides [12] or chalcogenides [13] have been synthesized and tested as catalysts, drug-nanocarriers, and for energy conversion and storage [14,15].

Silicon, silicon carbide, and silicon oxide nanoparticles have been reported to be promising materials in the areas of energy conversion, sensors, catalysis, and nanomedicine [16–21]. To date, a number of physical and chemical techniques have been reported for growing silicon nanoparticles, including pulsed laser ablation, plasma processing, ball milling, chemical vapor deposition, colloidal routes, and electrochemical etching [22]. In almost all of these methods, the used raw materials were bulk silicon or highly toxic Si precursors. By using silicon on insulator (SOI) film as substrate, Zywietz et al. [23] prepared silicon nanoparticles by the pulsed laser technique. In their experiments, a pulsed laser beam was focused on the surface substrate to produce a local melting, from which liquid silicon drops were formed. They proposed a mechanism for explaining the drop formation and its transformation into 160 nm-sized solid Si nanoparticles. Similarly, amorphous 3 μm Si microparticles were obtained by Garin et al. [24] using low pressure chemical vapor deposition (CVD) at temperature between 400 °C and 600 °C, using disilane as the Si source. In this case, disilane decomposed into a variety of H–Si species in the reactor bulk to reach supersaturation, and then nucleation occurred, followed by the growth of Si nanoparticles, which were collected on the polished surface of a Si (100) n-type wafer. Shavel et al. [25] produced 4–6 nm Si nanoparticles by a colloidal route using a mixture of $AlCl_3$–NaCl as molten salt solvent, some stabilizer, and silicon alkoxide precursor as (3-aminopropyl)triethoxysilane (APTES) or tetramethyl orthosilicate (TMOS). The synthesis was done at atmospheric pressure and 250 °C under Ar atmosphere.

On the other hand, composites of silicon nanoparticles embedded into carbon-derived materials are currently studied as the anode material of Li-ion batteries [26]. These nanocomposites have been prepared by electrospinning, and hydrothermal and CVD methods, and have been demonstrated to produce high-performance anodes in Li-ion battery research [27–29]. However, some degradation-related problems were detected, because nanosized silicon pulverizes, due to the stress promoted by the anode volume change during the cycling lithiation–delithiation processes [30]. Beyond the abovementioned ones, other applications for Si nanoparticles include solar cell manufacture [31] and biomarkers [32].

A promised application niche for silicon–carbon hybrid nanocomposites are as biological markers, since graphene oxide and silicon have a low toxicity compared with capped cysteamine or mercaptoundecanoic acid CdSe and CdTe quantum dots [33].

To our best knowledge, there are a few reports on the synthesis of carbon–silicon hybrid 1D nanocomposites. For instance, Chen et al. [28] reported the fabrication of embedded-silicon carbon nanowires, starting from Si nanoparticles that were first oxidized to form SiO_2–Si core–shell nanostructures. The obtained nanoparticles were dispersed in Pluronic F127, polyacrylonitrile, and *N,N*-dimethylformamide mixtures, and then electrospun into hybrid nanofibers having Si–SiO_2 nanoparticles, and the SiO_2 layer was then removed with HF.

This work reports a novel method to synthesize polycrystalline Si–SiC nanoparticles on a GO-coated silicon wafer surface. The present work derives from our previous research on the synthesis of graphite-encapsulated copper or copper oxide hybrid nanoparticles at the GO film/copper foil interface, which was formed by dripping water dispersed-GO on a copper foil surface. Here, it was observed that GO promotes the growth of Cu-based nanostructures at temperatures as low as 80 °C, and that their phase and morphology strongly depended on the processing temperature [34,35]. Based on these findings and considering that a solid melts starting from its surface, at temperatures below its nominal bulk melting point, we carried out experiments to produce graphite-encapsulated silicon nanoparticles at the surface of a silicon wafer by following a similar method as that used for making the graphite-coated copper-based nanoparticles.

2. Materials and Methods

2.1. Materials

Graphite flakes (+100 mesh), sodium nitrate (≥99%), potassium permanganate (≥99%), and hexane (≥99%) were obtained from Sigma-Aldrich (Toluca, Mexico). Sulfuric acid (95–98%) was purchased from Reproquifin (Mexico City, Mexico).Hydrogen peroxide (30%) and acetone (99.77%) were obtained from J.T. Baker (Mexico City, Mexico). Ethanol (99.5%) was purchased from Reasol (Mexico City, Mexico). All reagents were used as received without further purification.

2.2. Preparation of the CSNW

GO was prepared from natural graphite flakes by the modified Hummers' method, as previously reported by the authors [35]. A 1×1 cm^2 p-type silicon (Si) (100)-oriented wafer was used as the substrate. Before the GO film deposition, it was degreased by sequential sonication steps in hexane, acetone, and ethanol. GO films of different thickness were deposited by dripping ~0.02 mg mL^{-1} or ~2 mg mL^{-1} GO concentrated dispersions, to obtain samples named A and B, respectively. Finally, the samples were dried at 80 °C under nitrogen flux for 15 min to consolidate a film. The GO/Si samples were annealed in a quartz tube furnace at 1000 °C for 1 h under Ar atmosphere.

2.3. Characterization

The structural change of GO-silicon nanocomposites was assessed by Raman Spectroscopy using a WITec alpha 300 RA+ apparatus (WITec GmbH, Ulm, Germany), with an attached 50× objective lens and 532 nm laser excitation, recorded from deposited GO and reduced graphene oxide (rGO) films. Fourier transform infrared spectroscopy (FT-IR) was done on KBr-supported sample pellets using a Perkin-Elmer spectrum 100 (Perkin-Elmer Inc., Waltham, MA, USA). Film thickness was measured in a KLA-Tencor P-15 profilometer (KLA-Tencor Corp., Milpitas, CA, USA). The morphology was studied using a field-effect scanning electron microscope (FE-SEM) Zeiss Auriga (Carl Zeiss Microscopy GmbH., Jena, Germany) or Hitachi STEM-5500 (Hitachi Ltd., Tokio, Japan), and high-resolution transmission electron microscopes (HR-TEM) JEOL ARM 200F and JEOL 2010 (JEOL Ltd., Tokio, Japan).

3. Results and Discussion

3.1. Raman and FT-IR Characterization

According to Raman analysis, thermal processing at 1000 °C under an Ar atmosphere produced the simultaneous formation of the silicon nanoparticles or carbon–silicon nanowires and GO reduction. Similar results regarding the reduction level and the silicon–silicon carbide nanoparticles formation were obtained for both samples, A and B.

Figure 1 shows Raman spectrum (sample B) of (a) as-deposited GO and (b) one corresponding to GO–Si hybrid nanocomposite on the Si substrate. The as-deposited GO Raman spectrum exhibits the G and D bands at 1598 and 1357 cm^{-1}, respectively. As frequently reported, they originate from the in-plane vibrations of carbon sp^2 bonds (G band), and structural defects (D band) [36,37]. After the thermal process (Figure 1b), both bands changed in position and intensity, consistently with the GO reduction degree [38]. That is, the G and D peaks underwent a red-shift of about 13 and 10 cm^{-1}, respectively. The red shift of the G band suggests the reduction of graphene oxide and the recovery of the sp^2 domain, while the D band shift is associated with the size of the in-plane sp^2 domain [37,39]. Likewise, the increase in the I(D)/I(G) ratio from 0.91 to 1.21 indicates the formation of numerous aromatic domains of smaller overall size in graphene [40]. Note that Chan Lee et al. [41] reported the effective reduction of sprayed-GO films by Si.

Raman spectrum of rGO-Si nanocomposite (Figure 1b) displays the GO Raman bands and those belonging to silicon ~515 and 950 cm^{-1}. The sharp band at 515 cm^{-1} could be assigned to TO vibration of Si–Si bond [42], the red shift from the assigned value of bulk silicon (~520 cm^{-1}) could be attributed to the presence of nanocrystals in the nanowire [43]. The low intensity Raman band at 950 cm^{-1} appears to correspond to nanosized silicon, as reported by Meier et al. [44], however, the HR-TEM analysis of prepared material revealed that polycrystalline nanoparticles composed of Si and SiC were obtained.

On the other hand, the FT-IR characterization served to confirm the effective GO reduction as indicated by Raman spectroscopy. As-deposited GO (Figure 2a) shows absorption bands at 1117, 1201, 1427, 1622, and 3425 cm^{-1}, corresponding to C–O stretching vibrations, C–OH stretching vibrations, O–H deformation, C=C stretching (skeletal vibrations from unoxidized graphitic domains), and O–H stretching vibrations, respectively, all of them in good agreement with the wavenumber localization reported for GO prepared by the Hummers' method [45,46].

The FT-IR spectrum of GO–Si nanoparticle composite (Figure 2b) also suggests the reduction of thermally processed GO with the temperature, since all the absorption bands associated with the abovementioned oxygenated groups decreased in intensity. It was even observed that the thermal process led to the entire vanishing of some bands. In the FT-IR spectra, both of the silicon nanoparticles and of the carbon and silicon nanowires, the bands associated to Si–C or Si–O cannot be appreciated in Figure 2b, since the signals are screened, due to the core–shell structure [47].

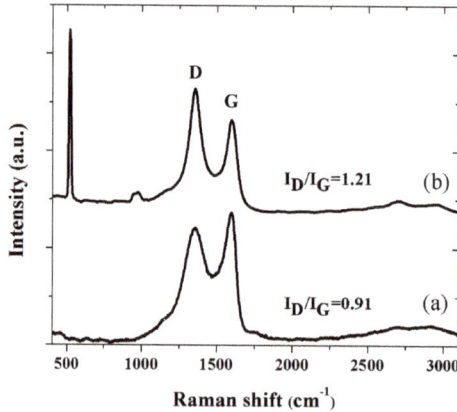

Figure 1. Raman spectra of as-deposited graphene oxide (GO) (**a**) and reduced graphene oxide (rGO)-Si hybrid nanocomposite (**b**). The latter displays Raman bands of Si (namely, the substrate) at 515 cm^{-1} and nanosized Si at 950 cm^{-1}.

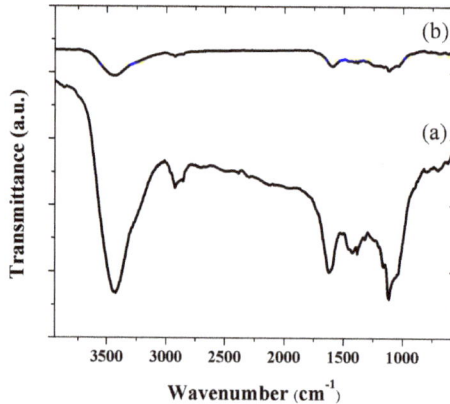

Figure 2. FT-IR spectra of (**a**) as-deposited GO and annealed sample at (**b**) 1000 °C for 1 h.

3.2. Morphology and Structure

After being prepared, the structure and morphology of samples A and B were examined by FE-SEM and HR-TEM. It was found that the thermal annealing led to the formation of silicon-derived nanoparticles at the GO/Si interface, and their morphology and mean particle size strongly depended on the GO film thickness. The HR-TEM analysis revealed that the nanoparticles are multiphase, comprising Si and SiC as dominant phases. The sample A (GO film thickness of around 100 nm) consists of Si–SiC nanoparticles of 20 nm mean size decorating the GO sheet surface (Figure 3a,b). On the other hand, in sample B (GO film thickness of 1 µm), Si–SiC nanoparticles (30 nm mean size) were self-assembled into worm-like hybrid nanostructures, as shown in the FE-SEM images at low- and high-magnification (Figure 3c,d, respectively). These hybrid rGO-Si–SiC worm-like nanostructures are mostly 10–100 nm in diameter with a length of several micrometers, covering a relatively large area of the Si wafer surface (note the magnitude scale of FE-SEM image in Figure 3d).

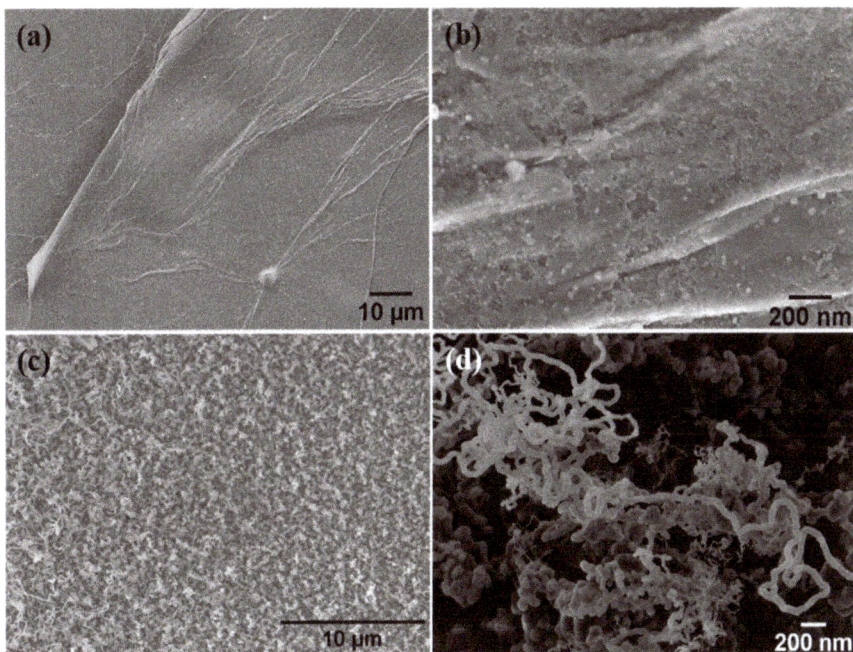

Figure 3. FE-SEM images of sample A (**a**,**b**) (nanoparticles decorating rGO sheets) and sample B (**c**,**d**) (the CSNW obtained from the thermal annealing of the GO at 1000 °C, using a Si wafer as support).

The TEM and HRTEM images in Figure 4 display the morphological features of samples A and B. Figure 4a corroborates the FE-SEM characterization result in that Si-based nanoparticles decorate the rGO sheets in the sample A. On the other hand, Figure 4c,d displays the striking morphology of sample B, which consists of different-sized nanoparticles that self-assembled to develop the worm-like structures, as those shown in Figure 3d. TEM images, Figure 4c,d, suggest that silicon-derived nanoparticles are embedded into 1D self-assembled rGO sheets. In Figure 4d, a magnified image of the framed zone in Figure 4b is shown. It provides a more detailed view of the branched characteristics of the worm-like nanostructures, as they are formed by rGO-supported individual nanoparticles.

The HR-TEM images of individual nanoparticles of samples A and B are shown in Figure 4b,e, respectively. It is seen that both are 30 nm diameter ball-shaped nanoparticles, apparently coated with an rGO shell. A careful determination of interplanar spacing revealed that various crystalline phases coexist in the formed nanoparticles, that is, they are multiphase polycrystalline entities; geometric shapes in Figure 4c were used to indicate the coexisting crystalline phases.

We report fast Fourier Transform (FTT) images in the inset of Figure 4b, and the areas selected in Figure 4c were used to estimate interplanar distances, which agree well with that reported for the spacing of (100), (002), (110), (111), and (131) planes of cubic SiC, (111), (113), and (353) of Si, and (110), (011), (302), and (300) of graphite (C). Note that, if they exist, we were unable to detect silicon oxide-derived phases.

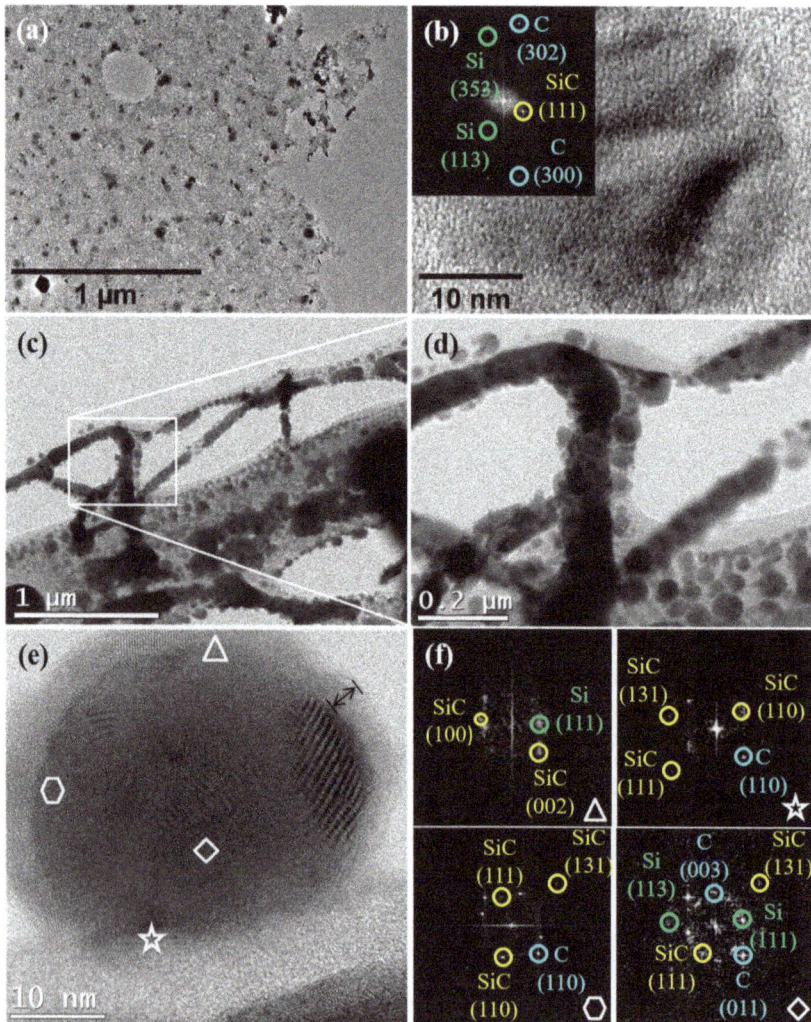

Figure 4. TEM and HR-TEM images of the sample A (**a**,**b**) and the sample B (**c**–**e**) after thermal annealing of GO at 1000 °C. (**a**) rGO sheets decorated by Si-based polydispersed nanoparticles. (**b**) HR-TEM of an irregularly shaped nanoparticle, where the inset corresponds to the FTT of (**b**). (**c**,**d**) Images of carbon–silicon-based nanowires (CSNW) which consist of Si-based nanoparticles embedded into 1D self-assembled rGO sheets. (**e**) HR-TEM of a nanoparticle, where it can be observed that it is polycrystalline. (**f**) The FTT of selected areas in (**e**).

3.3. Mechanism Proposed for the CSNW Formation

The morphological characterization by FE-SEM and HR-TEM suggests that polycrystalline Si–SiC nanoparticles arise as the primary building blocks to develop more complex nanostructures. Both the mean particle size and morphology are strongly associated to the GO film thickness. Indeed, experiments carried out on silicon wafer GO-free surface were unsuccessful in growing silicon nanoparticles. After being formed, they interact, leading to a self-assembly process that produces the worm-like structures.

To explain the silicon nanoparticle formation, we have assumed that they originate from melted silicon at the silicon surface through a nucleation process that starts at the high energy edges of GO sheets. Since the Si melting point is around 1414 °C [48], and the hybrid materials were made at 1000 °C, we proposed that impurities and structural defects at the silicon wafer surface might promote the melting of silicon surface (pre-melting effect) at temperatures lower than the nominal silicon melting point. Also, it is quite probable that GO has a certain participation in the melting of silicon surface, because recent reports suggest that GO could act as a high temperature reactor [49]. Thus, we believe that the synergy between the pre-melting effect and the ability of GO to absorb IR radiation [50] promotes melting of the silicon surface. Additional support for the silicon surface pre-melting at temperatures lower than its melting point is provided by the roughness change observed in thermal oxidation at the silicon wafer surface, at temperatures of 900–1000 °C [51–53].

Regarding the phase composition of nanoparticles, it is known that the temperature of GO pyrolysis strongly depends on GO sheet size. It is quite probable that the combined participation of both GO pyrolysis and CO_2 emission by GO reduction supply elemental carbon to react with silicon and, in so doing, form the observed silicon carbide. The formed nanoparticles are stabilized by the rGO network.

Accordingly, we propose that the observed worm-like hybrid nanostructures develop as illustrated in Figure 5.

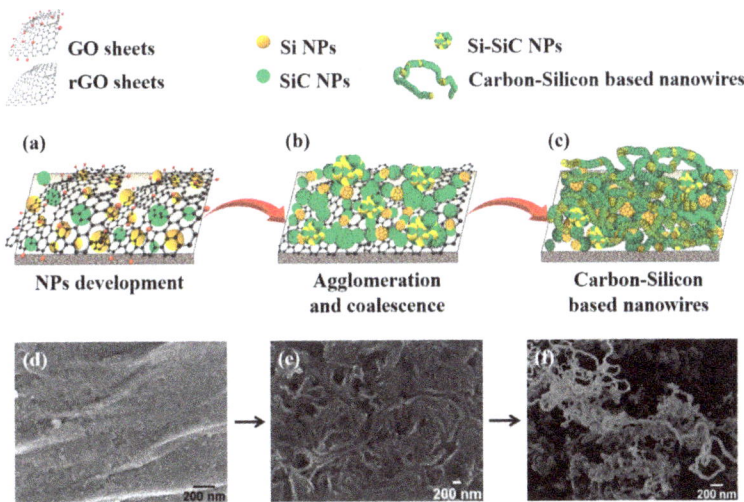

Figure 5. Scheme of the proposed mechanism for the carbon–silicon-based nanowires formation. (**a**) Si-based nanoparticle development at the interface GO and Si wafer. (**b**) nanoparticle agglomeration and coalescence into larger structures. (**c**) carbon-silicon based nanowires formation. FE-SEM images to support the proposed steps (**d**) development. (**e**) agglomeration and coalescence and (**f**) carbon-silicon based nanowires formation.

At first, small Si–SiC nanoparticles develop (Figure 5a), then, they diffuse across the melted silicon surface to interact and, finally, they coalesce to form larger nanostructures (Figure 5b). The worm-like structure originates from the GO-promoted self-assembly of individual Si, SiC, and Si–SiC polycrystalline nanoparticles, with the latter formed by the coalescence among Si and SiC (see Figure 5c), as suggested by HR-TEM (Figure 4). The FE-SEM images (Figure 5d–f) were added to support the proposed mechanism for the nanoparticle self-assembly.

4. Conclusions

In summary, we have presented the development of silicon–silicon carbide (Si–SiC) nanoparticles at the graphene oxide/silicon interface. Depending on the GO film thickness, the obtained nanoparticles disperse on the silicon surface, or they self-assemble into a few-micrometer 1D worm-like hybrid nanostructure. It was demonstrated that dispersed silicon nanoparticles develop on graphene oxide layers, whilst 1D nanostructures comprise GO-embedded self-assembled Si–SiC nanoparticles. It was assumed that the silicon surface melts, and that Si and SiC nanoparticles originated through a graphene oxide-induced nucleation of melted silicon. In addition, a mechanism for the formation of the (CSNW) was proposed.

Author Contributions: M.A.P.-G. and R.O.-A. participated in the conceptualization, investigation, and writing draft preparation, Y.M. and A.M.E.-R. participated in the methodology, validation, writing draft, and editing, J.M.-C. and J.S.-S. had participated in the data curation and writing review and editing, M.O.-L. participated in the conceptualization, supervision, and writing review and editing.

Funding: This research received no external funding.

Acknowledgments: This project is supported by the National Council of Science and Technology (CONACyT) scholarships Nos. 354945 (M.A.P.-G.) and 355201 (R.O.-A.). The authors are very thankful to Daniel Bahena Uribe for his valuable assistance for HR-TEM characterization and Jorge Roque De la Puente Ibarra for the FE-SEM images, both from LANE-Cinvestav. We are also grateful to Arturo Ponce-Pedraza from Physics Department and to the Kleberg Advanced Microscopy Center, University of Texas at San Antonio, for FE-SEM characterization. Finally, we would like to thank to Miguel Luna Arias from SEES, Cinvestav, for the profilometer measurements.

Conflicts of Interest: The authors declare no conflict of interest.

References

1. Ghosh, B.; Sarma, S.; Pontsho, M.; Ray, S.C. Tuning of magnetic behaviour in nitrogenated graphene oxide functionalized with iron oxide. *Diam. Relat. Mater.* **2018**, *89*, 35–42. [CrossRef]

2. Velasco-Soto, M.A.; Pérez-García, S.A.; Alvarez-Quintana, J.; Cao, Y.; Nyborg, L.; Licea-Jiménez, L. Selective band gap manipulation of graphene oxide by its reduction with mild reagents. *Carbon* **2015**, *93*, 967–973. [CrossRef]

3. Wang, R.; Lu, K.-Q.; Tang, Z.-R.; Xu, Y.-J. Recent progress in carbon quantum dots: Synthesis, properties and applications in photocatalysis. *J. Mater. Chem. A* **2017**, *5*, 3717–3734. [CrossRef]

4. Lucien, R.; Sunjie, Y.; Samuel, C.T.M.; Kevin, C.; Coletta, P.L.; Stephen, D.E. Morphological control of seedlessly-synthesized gold nanorods using binary surfactants. *Nanotechnology* **2018**, *29*, 135601.

5. Zhu, K.; Ju, Y.; Xu, J.; Yang, Z.; Gao, S.; Hou, Y. Magnetic Nanomaterials: Chemical Design, Synthesis, and Potential Applications. *Acc. Chem. Res.* **2018**, *51*, 404–413. [CrossRef] [PubMed]

6. Chen, F.; Hableel, G.; Zhao, E.R.; Jokerst, J.V. Multifunctional nanomedicine with silica: Role of silica in nanoparticles for theranostic, imaging, and drug monitoring. *J. Colloid Interface Sci.* **2018**, *521*, 261–279. [CrossRef] [PubMed]

7. Wu, Z.-S.; Zhou, G.; Yin, L.-C.; Ren, W.; Li, F.; Cheng, H.-M. Graphene/metal oxide composite electrode materials for energy storage. *Nano Energy* **2012**, *1*, 107–131. [CrossRef]

8. Thomas, V.J.; Ramaswamy, S. Application of Graphene and Graphene Compounds for Environmental Remediation. *Sci. Adv. Mater.* **2016**, *8*, 477–500. [CrossRef]

9. Bai, S.; Shen, X. Graphene-inorganic nanocomposites. *RSC Adv.* **2012**, *2*, 64–98. [CrossRef]

10. Huang, J.; Li, Y.; Jia, X.; Song, H. Preparation and tribological properties of core–shell $Fe_3O_4@C$ microspheres. *Tribol. Int.* **2019**, *129*, 427–435. [CrossRef]

11. Hou, L.; Zheng, H.; Cui, R.; Jiang, Y.; Li, Q.; Jiang, X.; Gao, J.; Gao, F. Silicon carbon nanohybrids with expandable space: A high-performance lithium battery anodes. *Microporous Mesoporous Mater.* **2019**, *275*, 42–49. [CrossRef]

12. Hong, W.G.; Kim, B.H.; Lee, S.M.; Yu, H.Y.; Yun, Y.J.; Jun, Y.; Lee, J.B.; Kim, H.J. Agent-free synthesis of graphene oxide/transition metal oxide composites and its application for hydrogen storage. *Int. J. Hydrogen Energy* **2012**, *37*, 7594–7599. [CrossRef]

13. Sarkar, S.; Howli, P.; Das, B.; Das, N.S.; Samanta, M.; Das, G.C.; Chattopadhyay, K.K. Novel Quaternary Chalcogenide/Reduced Graphene Oxide-Based Asymmetric Supercapacitor with High Energy Density. *ACS Appl. Mater. Interfaces* **2017**, *9*, 22652–22664. [CrossRef] [PubMed]

14. Piao, S.H.; Choi, H.J. Smart Graphene Oxide Based Composite Materials and their Electric and Magnetic Stimuli-response. *Procedia Eng.* **2017**, *171*, 64–70. [CrossRef]

15. Sun, S.; Zhou, A.; Gu, D.; Hu, X.; Li, Z.; Chen, L.; Ma, C.; Dong, L.; Yin, Y.; Chang, X. A general strategy for the synthesis of reduced graphene oxide-based composites. *Ceram. Int.* **2015**, *41*, 7661–7668. [CrossRef]

16. Ge, M.; Rong, J.; Fang, X.; Zhang, A.; Lu, Y.; Zhou, C. Scalable preparation of porous silicon nanoparticles and their application for lithium-ion battery anodes. *Nano Res.* **2013**, *6*, 174–181. [CrossRef]

17. Miyano, M.; Endo, S.; Takenouchi, H.; Nakamura, S.; Iwabuti, Y.; Shiino, O.; Nakanishi, T.; Hasegawa, Y. Novel Synthesis and Effective Surface Protection of Air-Stable Luminescent Silicon Nanoparticles. *J. Phys. Chem. C* **2014**, *118*, 19778–19784. [CrossRef]

18. Lal, S.; Caseley, E.A.; Hall, R.M.; Tipper, J.L. Biological Impact of Silicon Nitride for Orthopaedic Applications: Role of Particle Size, Surface Composition and Donor Variation. *Sci. Rep.* **2018**, *8*, 9109. [CrossRef] [PubMed]

19. Ege, D.; Kamali, A.R.; Boccaccini, A.R. Graphene Oxide/Polymer-Based Biomaterials. *Adv. Eng. Mater.* **2017**, *19*, 1700627. [CrossRef]

20. O'Farrell, N.; Houlton, A.; Horrocks, B.R. Silicon nanoparticles: Applications in cell biology and medicine. *Int. J. Nanomed.* **2006**, *1*, 451–472. [CrossRef]

21. Su, S.; He, Y.; Zhang, M.; Yang, K.; Song, S.; Zhang, X.; Fan, C.; Lee, S.-T. High-sensitivity pesticide detection via silicon nanowires-supported acetylcholinesterase-based electrochemical sensors. *Appl. Phys. Lett.* **2008**, *93*, 023113. [CrossRef]

22. Arunmetha, S.; Vinoth, M.; Srither, S.R.; Karthik, A.; Sridharpanday, M.; Suriyaprabha, R.; Manivasakan, P.; Rajendran, V. Study on Production of Silicon Nanoparticles from Quartz Sand for Hybrid Solar Cell Applications. *J. Electron. Mater.* **2018**, *47*, 493–502. [CrossRef]

23. Zywietz, U.; Evlyukhin, A.B.; Reinhardt, C.; Chichkov, B.N. Laser printing of silicon nanoparticles with resonant optical electric and magnetic responses. *Nat. Commun.* **2014**, *5*, 3402. [CrossRef] [PubMed]

24. Garín, M.; Fenollosa, R.; Alcubilla, R.; Shi, L.; Marsal, L.F.; Meseguer, F. All-silicon spherical-Mie-resonator photodiode with spectral response in the infrared region. *Nat. Commun.* **2014**, *5*, 3440. [CrossRef] [PubMed]

25. Shavel, A.; Guerrini, L.; Alvarez-Puebla, R.A. Colloidal synthesis of silicon nanoparticles in molten salts. *Nanoscale* **2017**, *9*, 8157–8163. [CrossRef] [PubMed]

26. Shen, C.; Barrios, E.; Zhai, L. Bulk Polymer-Derived Ceramic Composites of Graphene Oxide. *ACS Omega* **2018**, *3*, 4006–4016. [CrossRef]

27. Jeong, M.-G.; Du, H.L.; Islam, M.; Lee, J.K.; Sun, Y.-K.; Jung, H.-G. Self-Rearrangement of Silicon Nanoparticles Embedded in Micro-Carbon Sphere Framework for High-Energy and Long-Life Lithium-Ion Batteries. *Nano Lett.* **2017**, *17*, 5600–5606. [CrossRef] [PubMed]

28. Chen, Y.; Hu, Y.; Shen, Z.; Chen, R.; He, X.; Zhang, X.; Li, Y.; Wu, K. Hollow core–shell structured silicon@carbon nanoparticles embed in carbon nanofibers as binder-free anodes for lithium-ion batteries. *J. Power Sources* **2017**, *342*, 467–475. [CrossRef]

29. Huang, R.; Fan, X.; Shen, W.; Zhu, J. Carbon-coated silicon nanowire array films for high-performance lithium-ion battery anodes. *Appl. Phys. Lett.* **2009**, *95*, 133119. [CrossRef]

30. Wang, J.; Fan, F.; Liu, Y.; Jungjohann, K.L.; Lee, S.W.; Mao, S.X.; Liu, X.; Zhu, T. Structural Evolution and Pulverization of Tin Nanoparticles during Lithiation-Delithiation Cycling. *J. Electrochem. Soc.* **2014**, *161*, F3019–F3024. [CrossRef]

31. Gribov, B.G.; Zinov'ev, K.V.; Kalashnik, O.N.; Gerasimenko, N.N.; Smirnov, D.I.; Sukhanov, V.N.; Kononov, N.N.; Dorofeev, S.G. Production of Silicon Nanoparticles for Use in Solar Cells. *Semiconductors* **2017**, *51*, 1675–1680. [CrossRef]

32. Zyuzin, M.V.; Baranov, D.G.; Escudero, A.; Chakraborty, I.; Tsypkin, A.; Ushakova, E.V.; Kraus, F.; Parak, W.J.; Makarov, S.V. Photoluminescence quenching of dye molecules near a resonant silicon nanoparticle. *Sci. Rep.* **2018**, *8*, 6107. [CrossRef] [PubMed]

33. Hardman, R. A Toxicologic Review of Quantum Dots: Toxicity Depends on Physicochemical and Environmental Factors. *Environ. Health Perspect.* **2006**, *114*, 165–172. [CrossRef] [PubMed]

34. Ortega-Amaya, R.; Matsumoto, Y.; Flores-Conde, A.; Pérez-Guzmán, M.A.; Ortega-López, M. In situ formation of rGO quantum dots during GO reduction via interaction with citric acid in aqueous medium. *Mater. Res. Express* **2016**, *3*, 105601. [CrossRef]

35. Ortega-Amaya, R.; Matsumoto, Y.; Pérez-Guzmán, M.A.; Ortega-López, M. In situ synthesis of Cu_2O and Cu nanoparticles during the thermal reduction of copper foil-supported graphene oxide. *J. Nanopart. Res.* **2015**, *17*, 397. [CrossRef]

36. Sobon, G.; Sotor, J.; Jagiello, J.; Kozinski, R.; Zdrojek, M.; Holdynski, M.; Paletko, P.; Boguslawski, J.; Lipinska, L.; Abramski, K.M. Graphene Oxide vs. Reduced Graphene Oxide as saturable absorbers for Er-doped passively mode-locked fiber laser. *Opt. Express* **2012**, *20*, 19463–19473. [CrossRef] [PubMed]

37. Chen, W.; Yan, L. Preparation of graphene by a low-temperature thermal reduction at atmosphere pressure. *Nanoscale* **2010**, *2*, 559–563. [CrossRef] [PubMed]

38. Ji, Z.; Shen, X.; Li, M.; Zhou, H.; Zhu, G.; Chen, K. Synthesis of reduced graphene oxide/CeO_2 nanocomposites and their photocatalytic properties. *Nanotechnology* **2013**, *24*, 115603. [CrossRef] [PubMed]

39. Zhang, Y.; Ma, H.-L.; Zhang, Q.; Peng, J.; Li, J.; Zhai, M.; Yu, Z.-Z. Facile synthesis of well-dispersed graphene by [gamma]-ray induced reduction of graphene oxide. *J. Mater. Chem.* **2012**, *22*, 13064–13069. [CrossRef]

40. Stankovich, S.; Dikin, D.A.; Piner, R.D.; Kohlhaas, K.A.; Kleinhammes, A.; Jia, Y.; Wu, Y.; Nguyen, S.T.; Ruoff, R.S. Synthesis of graphene-based nanosheets via chemical reduction of exfoliated graphite oxide. *Carbon* **2007**, *45*, 1558–1565. [CrossRef]

41. Chan Lee, S.; Some, S.; Wook Kim, S.; Jun Kim, S.; Seo, J.; Lee, J.; Lee, T.; Ahn, J.-H.; Choi, H.-J.; Chan Jun, S. Efficient Direct Reduction of Graphene Oxide by Silicon Substrate. *Sci. Rep.* **2015**, *5*, 12306. [CrossRef] [PubMed]

42. Kole, A.; Chaudhuri, P. Growth of silicon quantum dots by oxidation of the silicon nanocrystals embedded within silicon carbide matrix. *AIP Adv.* **2014**, *4*, 107106. [CrossRef]

43. Parul, S.; Anguita, J.V.; Stolojan, V.; Henley, S.J.; Silva, S.R.P. The growth of silica and silica-clad nanowires using a solid-state reaction mechanism on Ti, Ni and SiO_2 layers. *Nanotechnology* **2010**, *21*, 295603.

44. Meier, C.; Lüttjohann, S.; Kravets, V.G.; Nienhaus, H.; Lorke, A.; Wiggers, H. Raman properties of silicon nanoparticles. *Phys. E Low-Dimens. Syst. Nanostruct.* **2006**, *32*, 155–158. [CrossRef]

45. Dong, J.; Liu, W.; Li, H.; Su, X.; Tang, X.; Uher, C. In situ synthesis and thermoelectric properties of PbTe-graphene nanocomposites by utilizing a facile and novel wet chemical method. *J. Mater. Chem. A* **2013**, *1*, 12503–12511. [CrossRef]

46. Shen, J.; Yan, B.; Shi, M.; Ma, H.; Li, N.; Ye, M. One step hydrothermal synthesis of TiO_2-reduced graphene oxide sheets. *J. Mater. Chem.* **2011**, *21*, 3415–3421. [CrossRef]

47. Necmi, S.; Tülay, S.; Şeyda, H.; Yasemin, Ç. Annealing effects on the properties of copper oxide thin films prepared by chemical deposition. *Semicond. Sci. Technol.* **2005**, *20*, 398.

48. Durand, F.; Duby, J.C. Carbon solubility in solid and liquid silicon—A review with reference to eutectic equilibrium. *J. Phase Equilib.* **1999**, *20*, 61. [CrossRef]

49. Barman, B.K.; Nanda, K.K. Ultrafast-Versatile-Domestic-Microwave-Oven Based Graphene Oxide Reactor for the Synthesis of Highly Efficient Graphene Based Hybrid Electrocatalysts. *ACS Sustain. Chem. Eng.* **2018**, *6*, 4037–4045. [CrossRef]

50. Bae, J.J.; Yoon, J.H.; Jeong, S.; Moon, B.H.; Han, J.T.; Jeong, H.J.; Lee, G.-W.; Hwang, H.R.; Lee, Y.H.; Jeong, S.Y.; et al. Sensitive photo-thermal response of graphene oxide for mid-infrared detection. *Nanoscale* **2015**, *7*, 15695–15700. [CrossRef] [PubMed]

51. Araki, K.; Takeda, R.; Sudo, H.; Izunome, K.; Zhao, X. Dependence of Atomic-Scale Si(110) Surface Roughness on Hydrogen Introduction Temperature after High-Temperature Ar Annealing. *J. Surf. Eng. Mater. Adv. Technol.* **2014**, *4*, 249–256. [CrossRef]

52. Shive, L.W.; Gilmore, B.L. Impact of Thermal Processing on Silicon Wafer Surface Roughness. *ECS Trans.* **2008**, *16*, 401–405.

53. Shi, Z.; Shao, S.; Wang, Y. Improved the Surface Roughness of Silicon Nanophotonic Devices by Thermal Oxidation Method. *J. Phys. Conf. Ser.* **2011**, *276*, 012087. [CrossRef]

![nanomaterials logo] *nanomaterials*

MDPI

Article

3C-SiC Nanowires In-Situ Modified Carbon/Carbon Composites and Their Effect on Mechanical and Thermal Properties

Hongjiao Lin [1], Hejun Li [1,*], Qingliang Shen [1], Xiaohong Shi [1,*], Tao Feng [1,2] and Lingjun Guo [1]

[1]   State Key Laboratory of Solidification Processing, Northwestern Polytechnical University, Xi'an 710072, China; lin.hong.jiao@163.com (H.L.); shenqingliang@mail.nwpu.edu.cn (Q.S.); fengtao@nwpu.edu.cn (T.F.); guolingjun@nwpu.edu.cn (L.G.)

[2]   School of Mechanics, Civil Engineering & Architecture, Northwestern Polytechnical University, Xi'an 710072, China

*   Correspondence: lihejun@nwpu.edu.cn (H.L.); npusxh@nwpu.edu.cn (X.S.); Tel.: +86-29-8849-2272 (H.L.); +86-29-8849-2711 (X.S.); Fax: +86-29-8849-2642 (H.L.); +86-29-8849-5764 (X.S.)

Received: 8 October 2018; Accepted: 30 October 2018; Published: 1 November 2018

Abstract: An in-situ, catalyst-free method for synthesizing 3C-SiC ceramic nanowires (SiCNWs) inside carbon–carbon (C/C) composites was successfully achieved. Obtained samples in different stages were characterized by X-ray diffraction (XRD), scanning electron microscopy (SEM), and Raman scattering spectroscopy. Results demonstrated that the combination of sol-gel impregnation and carbothermal reduction was an efficient method for in-situ SiCNW synthesis, inside C/C composites. Thermal properties and mechanical behaviors—including out-of-plane and in-plane compressive strengths, as well as interlaminar shear strength (ILLS) of SiCNW modified C/C composites—were investigated. By introducing SiCNWs, the initial oxidation temperature of C/C was increased remarkably. Meanwhile, out-of-plane and in-plane compressive strengths, as well as interlaminar shear strength (ILLS) of C/C composites were increased by 249.3%, 109.2%, and 190.0%, respectively. This significant improvement resulted from simultaneous reinforcement between the fiber/matrix (F/M) and matrix/matrix (M/M) interfaces, based on analysis of the fracture mechanism.

Keywords: SiC nanowires; C/C composites; in-situ growth; mechanical properties

1. Introduction

Carbon–carbon (C/C) composites, composed of carbon fibers and a carbon matrix, are important thermal-structural materials. C/C composites can retain their strength and stiffness under ultra-high temperatures, and they exhibit unique thermal properties, such as high heat capacity, high thermal conductivity, low thermal expansion, and good thermal shock resistance. Therefore, C/C composites are widely applied in aerospace engineering and used in industrial devices [1–3]. Mechanical and thermal properties of C/C composites mainly depend on their carbon microstructures. Generally, the carbon matrix is deposited inside the carbon fiber preform by chemical vapor infiltration (CVI), to form C/C composites. In addition, the process of infiltrating the carbon matrix greatly affects the interface in yielded composites. Bridge interfaces for both stress and heat transfer significantly impact the mechanical and thermal performance of C/C composites [4–6], as well as impact the mechanical and thermal performances of C/C composites.

Mechanical and thermal properties are key performance indicators for assessing high-temperature structure materials [7]. As for C/C composites, mechanical and thermal properties mainly depend on the microstructure of the carbon materials. An interface is the bridge, through which load and heat can be transferred in composites [8,9]. There are mainly two kinds of interfaces in C/C composites: fiber/matrix (F/M) interfaces—interfaces between the fiber and its surrounding matrix—and matrix/matrix (M/M) interfaces, which result from the layered feature of the matrix and refers to interfaces between the layered matrix. The poor bonding strength of both interfaces leads to low interlaminar shear strength and compressive strength of the composites. In recent years, numerous investigations have been carried out to modify these interfaces, in order to improve both the mechanical and thermal performance of C/C composites [10–12].

Silicon carbide nanowires (SiCNWs) have attracted much interest because of their excellent elasticity and strength, which are much better than those of SiC ceramic bulks and whiskers. Therefore, SiCNWs have been proven to have outstanding secondary reinforcing phases in ceramic, metals and polymer matrix composites [13–16]. However, the application of SiCNWs inside of C/C composites has been rarely studied. Up to now, the most commonly applied method employed has been the well-known, vapor-liquid-solid (VLS) growth mechanism, which usually uses a metal as the catalyst, due to its unique properties [17]. However, it is very difficult to remove the residual metal catalyst from SiC nanowires and it always acts as a contamination, resulting in the degradation of their electronic and mechanical properties, under harsh environmental conditions [18,19]. Especially when SiC nanowires are applied as the reinforced phase in C/C composites at high temperatures, the metal catalyst will cause devastating side effects. Catalysts can accelerate the oxidation process, which can corrode fibers. At the same time, because of the complexity of the internal structure of C/C composites, it is difficult to perform in-situ synthesis of nanowires. Hence, a greater enhancement of the embedded reinforcement phases in C/C composites should be carried out. Technological challenges affecting C/C composites are summarized below: (a) metal catalysts should be avoided; (b) a porous material with as many growth spaces as the growth medium is needed; (c) the reinforcement phase should be dispersed uniformly inside the matrix.

In the present work, we demonstrated a feasible way to achieve in-situ, catalyst-free growth of SiCNWs in C/C composites. Porous carbon felt was used as the growth medium, whereas silica xerogel and pyrolytic carbon were used as the source materials. Finally, SiCNW-C/C composites were obtained through densification, using the TGCVI (Thermal Gradient Chemical Vapor Infiltration) method. The morphology, crystal microstructure, composition, and growth mechanism of SiCNWs were studied. As for the SiCNW-C/C composites, mechanical properties—including out-of-plane and in-plane compressive strengths, as well as interlaminar shear strength (ILLS), were also tested. Furthermore, we analyzed the reinforcement mechanism of C/C composites with SEM images of the fracture surfaces, which demonstrated the excellent mechanical properties of SiCNWs during the reinforcement phase.

2. Experimental Section

2.1. Preparation of SiCNW-C/C Composites

Commercially available tetraethoxysilane (TEOS), ethanol absolute (EtOH, AR; Hongyan Reagent Factory, Tianjin, China), deionized water (H_2O, AR), hydrochloric acid (HCl, AR; Hongyan Reagent Factory, Tianjin, China), and 2D needled carbon felt (density: 0.54–0.56 g/cm^3, fiber diameter: 6–7 μm; Tianniao, Jiangsu, China) were used as raw materials. SiO_2 sol was prepared by the method reported in our previous work [20]. Figure 1 shows the schematic illustration for the fabrication of C/C composites, with in-situ grown SiCNWs. Firstly, 2D needled carbon fiber felts (120 mm × 80 mm × 70 mm) were impregnated with SiO_2 sol to obtain carbon felt/silica xerogel. Then the carbon felt/silica xerogel was placed at the center of a tube furnace. The heating rate was 10 °C/min and the flow rate of Ar gas was controlled at 1.5–1.6 L/h. When the temperature reached 1050 °C, it was maintained for 2 h. After that, the furnace was naturally cooled to room temperature and SiOC-C/C composite preforms were obtained through the ICVI (Isothermal Chemical Vapor Infiltration) reaction process.

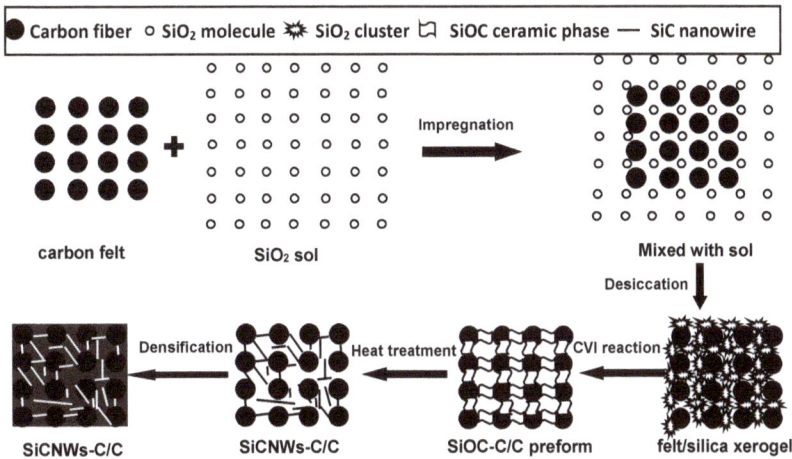

Figure 1. Schematic diagram of the formation of silicon carbide nanowires with carbon-carbon (SiCNW-C/C) composites. CVI = chemical vapor infiltration.

In order to obtain the SiCNWs, SiOC-C/C preforms were heat treated in a graphite furnace, at 1550 °C for 2 h, under an Ar atmosphere and with a steady heating rate of 20 °C/min. After the heat treatment, the SiOC was converted into many cross-linked SiC nanowires, as shown in Figure 2. The last step was the densification of the preforms by TGCVI, with natural gas as the carbon source, nitrogen as the dilute, and a carrier gas at 950 °C. After 140 h of densification, the final density of the SiCNW-C/C composite was about 1.80 g/cm^3.

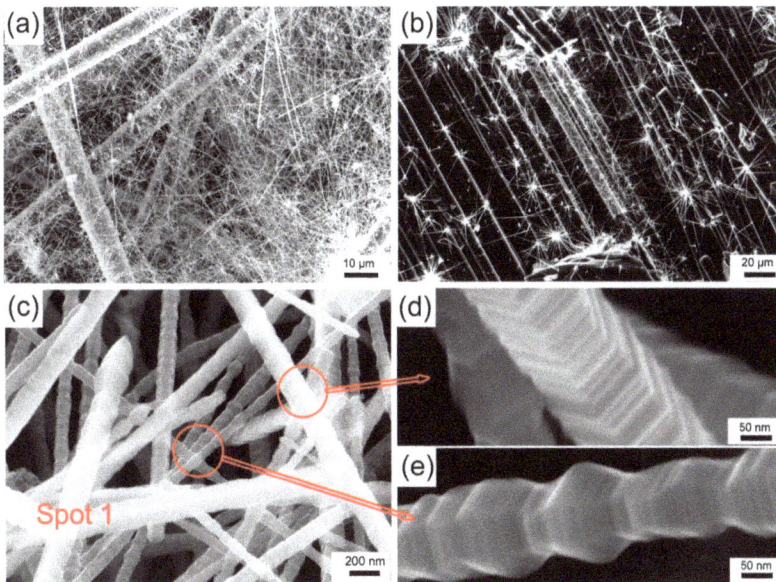

Figure 2. *Cont.*

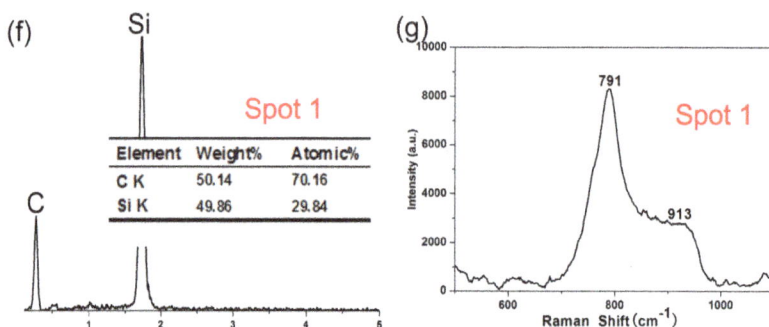

Figure 2. (**a**) Low-magnification of the needling felt layer, (**b**) low-magnification of the non-woven layer, (**c**) high-magnification, (**d**) columnar, and (**e**) beaded SEM images of the SiCNWs, (**f**) EDS of the SiCNWs, (**g**) and Raman spectra of the SiCNWs.

2.2. Characterization

Scanning electron microscopy (SEM) images were taken on a field emission scanning electron microscopy (FESEM, ZEISS-SUPRA 55, Jena, Germany), equipped with energy dispersive X-ray spectroscopy (EDS). The chemical composition and phase identification of the synthesized felt/silica xerogel, SiOC-C/C preform, and SiOC-C/C and SiCNW-C/C composites were analyzed using Raman scattering spectroscopy (Raman, Renishaw inVia, Gloucestershire, England) and X-ray diffraction (XRD; PANalytical X'Pert Pro, Almelo, The Netherlands), with Cu Kα radiation ($\lambda = 1.5418°A$).

2.3. Test of Mechanical Properties

Differential thermal analysis and thermal gravimetry were carried out on a TGA/SDTA851 thermoanalyzer (Mettler Toledo, Zurich, Switzerland), within an air atmosphere, at a heating rate of 10 °C/min. Compressive strength (out-of-plane and in-of-plane, ASTM D3410) and interlaminar shear strength (ILSS; ASTM D5379)—as well as the compressive shearing of the composites before and after modification by the SiCNWs—were tested to demonstrate the effect of SiCNWs in increasing the C/C composite's mechanical properties. The samples prepared for tests, with a loading rate of 0.5 mm/min at room temperature, were placed in square grids with the size of 10 mm × 10 mm × 10 mm. All these samples were tested, using an electronic universal testing machine (Shenzhen SANS Testing Machine Co., Ltd., SANS CMT5304-30, Shenzhen, China). Final results were achieved by computing means of at least five samples. Both the compressive strength and shear strength were calculated by $\sigma = P/S$, where σ means the calculated mechanical strength, P denotes the compressive force, and S signifies the area of force.

3. Results and Discussion

3.1. Morphology of SiCNWs

The as-obtained SiCNWs were characterized by SEM to identify their morphology. The non-woven layer and needling felt layer of the SiCNW-C/C preform could be observed from the low magnification SEM image of the SiCNWs. As shown in Figure 2a,b, there was a large number of nanowires randomly distributed on the surface of the carbon fibers, whose lengths varied from several to hundreds of micrometers. Compared to the needling felt layer, SiCNWs in the non-woven layer were relatively few and radiated from the carbon fiber to the surrounding stretch. This was because there was more reaction space and Si source in the needling felt layer. Figure 2c indicates that mainly two types of SiCNWs existed. Corresponding high-magnification SEM images in Figure 2d,e clearly reveal the shape of the nanowires, with a diameter of approximately 100–200 nm, and were columnar and beaded, respectively. The specific elemental composition of the as-prepared sample, taken from Spot 1 (in red) in Figure 2c was further investigated using EDS, as displayed in the inset in Figure 2f. Due to the influence of the carbon matrix, the atomic ratio of Si and C was about 1:3. The Raman spectra (Figure 2g) show a peak at around 791 cm^{-1} and a broad peak in Spot 1 (in red), which were attributed to the 3C-SiC nanowires [21].

3.2. Phase and Structure Characterization

Figure 3a–c shows the X-ray diffraction pattern of the carbon fiber felt, felt/silica xerogel, and SiOC-C/C preform. One main symmetrical diffraction peak at 25.9° was observed, which was assigned to the diffraction plane (002) of the sp^2 carbon materials [22,23]. Figure 3d shows the diffraction pattern of the fiber felt, after the growth of the SiCNWs. It is characterized by the extra four diffraction peaks, in comparison to the other samples. Intensities of the main peaks at 35.7°, 60.1°, and 71.2° all match the 3C-SiC nanowires (JCPDS Card no. 29-1129) which were attributed to the diffraction of the <beta>-SiC (111), (220), and (311) planes [24]. Additionally, the low-intensity peak (SF) at 33.48° on the left shoulder of (111) peak was typically observed in the XRD spectra of the 3C-SiC nanowires, which was usually ascribed to stacking faults within the crystals [25].

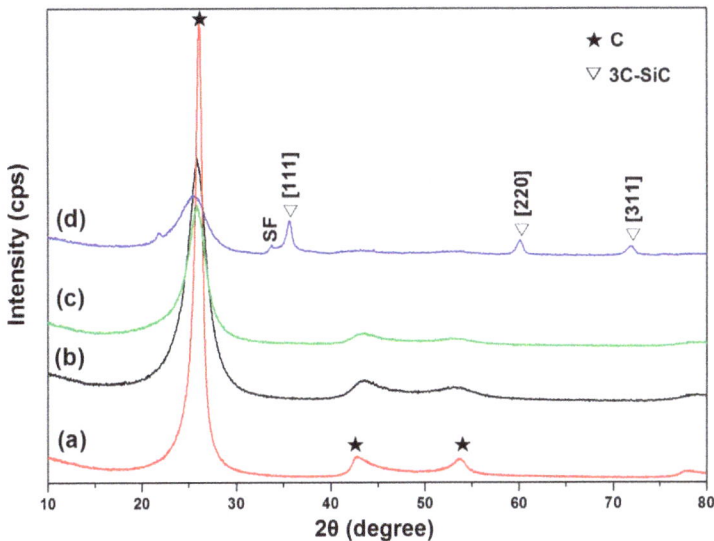

Figure 3. X-ray diffraction patterns for composites at different stages: (**a**) The carbon felt, (**b**) felt/silica xerogel, (**c**) SiOC-C/C preform, and (**d**) SiCNW-C/C perform.

Typical Raman spectra of specimens at different stages are shown in Figure 4. There were four spectra of different stage samples which contained: Carbon felt (a), felt/silica xerogel (b), the SiOC-C/C preform (c), and (d) the SiCNW-C/C preform. The bands located near 1340 cm^{-1} (D band), 1580 cm^{-1} (G band), 2690cm^{-1} (2D band), and 2930 cm^{-1} (D+G band) can be clearly observed in the four spectra. According to previous research [26,27], the peak at 1340 cm^{-1} and 1580 cm^{-1} are usually associated with in-plane vibrations and in all sp^2 bonded carbon atoms. The weak peak near 2690 cm^{-1} was due to the overtone of the D band and the broad peak, at about 2930 cm^{-1}, was attributed to the sum of the D and G bands. A peak at around 790 cm^{-1} and a broad peak from 914–970 cm^{-1} were observed in Figure 4d, which were attributed to the zone center transverse optical (TO) phonon modes and the longitudinal optic (LO) phonon mode of the 3C-SiC nanowires [28]. We noted that the LO (G) phonon line showed a low number compared to other SiC materials [29,30]. One possible reason for this broadening was the size confinement effect: When the size of a crystal is reduced to a nano scale—such as in nanowires—phonons can be confined in the space by crystal boundaries or defects. As it is the reactive material source of SiC, SiO_2 may also have led to a widening of the Raman signals [31].

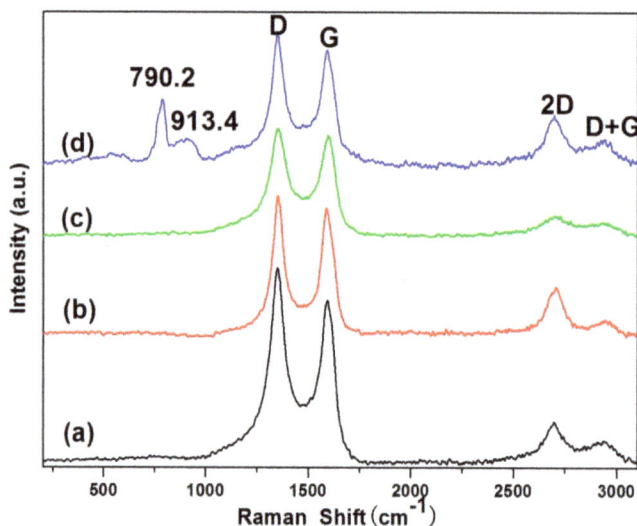

Figure 4. Raman spectra for composites at different stages: (**a**) The carbon felt, (**b**) felt/silica xerogel, (**c**) SiOC-C/C preform, and (**d**) SiCNW-C/C perform.

3.3. Reactivity of SiCNWs

On the basis of the experimental results above, we propose a possible mechanism for the sol-gel process, the CVI reaction process, and the heat treatment process, shown in Scheme 1. In the sol-gel process, TEOS was firstly hydrolyzed into silanol (Reaction 1), then two silanol molecules lost one H_2O molecule due to an elimination reaction (Reaction 2) [32,33]. Finally, $(SiO_4)_x$ could be obtained because of the interaction between the silanol molecules (Reaction 3). In the CVI reaction process, C vapor was obtained from decomposing CH_4, as indicated in Reaction 4. When the C vapor seeped into the porous $(SiO_4)_x$, Reactions 5, 6, 7, and 8 could occur, from the view point of thermodynamics. After that, the cross-linking reactions [34] between products of SiC_4, Si_3O, SiC_2O_2, and $SiCO_3$, reacted. Finally, the porous SiOC material could be received. As no catalyst was used during the preparing procedure, no droplets were observed at the tips or over the surfaces of the SiCNWs by the SEM and TEM characterizations. Therefore, we propose a catalyst-free vs. growth mechanism for the growth of SiCNWs [25,35,36]. As shown in the heat treatment process, thermal SiOC can readily decompose into free SiO groups and C (Reaction 10). When the pressure of SiO and C vapor was elevated to supersaturating conditions, the carbothermal reduction reaction involving two key reactions (Reactions 11 and 12) occurred inside the carbon felt, which led to the growth of SiCNWs. The initial SiC embryos could be formed by heterogeneous nucleation of the SiO gas with carbon particles, according to Reaction 11. As SiO and CO deposited on the tip of the growing SiC embryos, growth of SiCNWs gradually proceeded by the gas-phase reaction (Reaction 12) [23]. Another important aspect was that the porous characteristic of carbon felt and SiOC not only provided a wide growth space but also increased the reaction area.

(1) The sol-gel process (the source of Si)

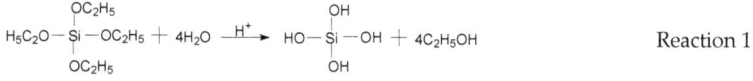

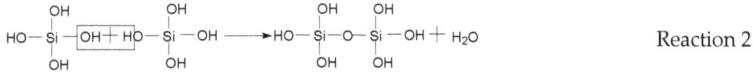

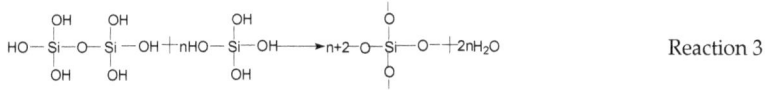

$$H_5C_2O-\underset{\underset{OC_2H_5}{|}}{\overset{\overset{OC_2H_5}{|}}{Si}}-OC_2H_5 + 4H_2O \xrightarrow{H^+} HO-\underset{\underset{OH}{|}}{\overset{\overset{OH}{|}}{Si}}-OH + 4C_2H_5OH$$

Reaction 1

$$HO-\underset{\underset{OH}{|}}{\overset{\overset{OH}{|}}{Si}}-\boxed{OH + HO}-\underset{\underset{OH}{|}}{\overset{\overset{OH}{|}}{Si}}-OH \longrightarrow HO-\underset{\underset{OH}{|}}{\overset{\overset{OH}{|}}{Si}}-O-\underset{\underset{OH}{|}}{\overset{\overset{OH}{|}}{Si}}-OH + H_2O$$

Reaction 2

$$HO-\underset{\underset{OH}{|}}{\overset{\overset{OH}{|}}{Si}}-O-\underset{\underset{OH}{|}}{\overset{\overset{OH}{|}}{Si}}-OH + nHO-\underset{\underset{OH}{|}}{\overset{\overset{OH}{|}}{Si}}-OH \longrightarrow n+2-O-\underset{\underset{O}{|}}{\overset{\overset{O}{|}}{Si}}-O- + 2nH_2O$$

Reaction 3

(2) The CVI reaction process (the source of C and the formation of SiOC)

$$CH_4 \longrightarrow C + 2H_2$$

Reaction 4

$$-O-\underset{\underset{O}{|}}{\overset{\overset{O}{|}}{Si}}-O- + 8C \longrightarrow SiC_4 + 4CO$$

Reaction 5

$$-O-\underset{\underset{O}{|}}{\overset{\overset{O}{|}}{Si}}-O- + 6C \longrightarrow SiC_3O + 3CO$$

Reaction 6

$$-O-\underset{\underset{O}{|}}{\overset{\overset{O}{|}}{Si}}-O- + 4C \longrightarrow SiC_2O + 2CO$$

Reaction 7

$$-O-\underset{\underset{O}{|}}{\overset{\overset{O}{|}}{Si}}-O- + 2C \longrightarrow SiCO_3 + CO$$

Reaction 8

$$SiC_4, SiC_3O, SiC_2O, SiCO_3 \xrightarrow{coexistence} SiOC \ \ ceramics$$

Reaction 9

(3) The heat treatment process (the formation of SiCNWs)

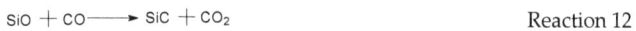

$$SiOC \xrightarrow{\triangle} SiO + C$$

Reaction 10

$$SiO + C \longrightarrow SiC + CO$$

Reaction 11

$$SiO + CO \longrightarrow SiC + CO_2$$

Reaction 12

Scheme 1. The possible reaction mechanism.

3.4. Thermal Analysis

TGA (thermo-gravimetric analysis) and SDTA (synchronous differential thermal analysis) curves of the C/C and SiCNW-C/C composites—shown in Figure 5a,b—were determined to investigate thermal stability. From Figure 5a, it can be clearly seen that initial oxidation decomposition temperatures of SiCNW-C/C composites were higher than those of pure C/C, whose oxidation started at 750 °C and ended at 1050 °C. Compared with C/C composites, the initial and end temperatures differed by about 150 °C and 90 °C. Undecomposed parts of the SiCNW-C/C and C/C composites were 15% and 0%, respectively, meaning C/C composites were nearly completely oxidized and SiCNW-C/C composites were not. From the SDTA curves, shown in Figure 5b, both C/C and SiCNW-C/C composites had obvious exothermic peaks, but the heat released from the C/C oxidation reaction was significantly more than that of the SiCNW-C/C composite. Therefore, the addition of SiCNWs could decrease the heat of the reaction to some extent.

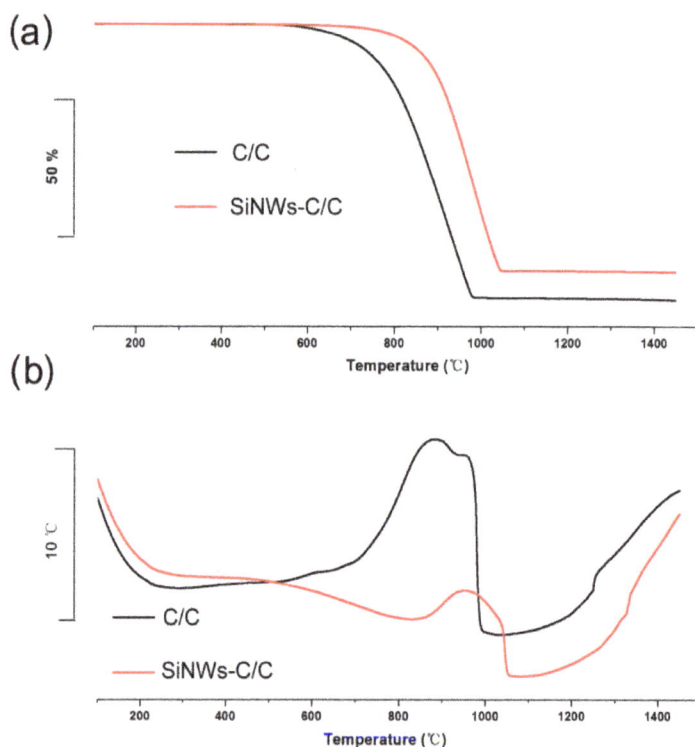

Figure 5. TGA (thermo-gravimetric analysis) (**a**) and SDTA (synchronous differential thermal analysis) (**b**) curves of the C/C and SiCNW-C/C composite samples.

3.5. Mechanical Properties

As shown in Figure 6 and Table 1, mechanical property tests indicated that SiCNW-C/C composites had significant improvements, including out-of-plane and in-plane compressive strengths, as well as interlaminar shear strength (ILSS) of 227%, 109%, and 190%, respectively. Compared with C/C composites, SiCNW-C/C composites also had 49% and 111% improvements in the out-of-plane and in-plane compressive moduli.

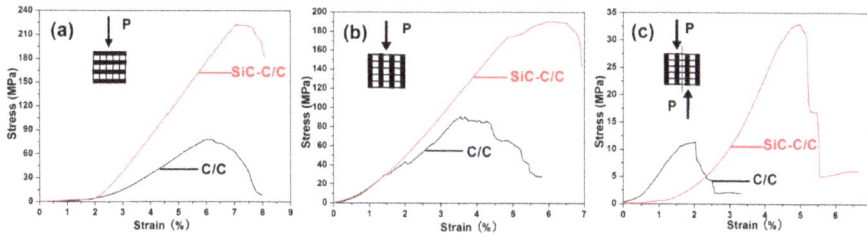

Figure 6. Mechanical properties of C/C and SiCNW-C/C composites: (**a**) Out-of-plane and (**b**) in-plane compressive strengths and (**c**) interlaminar shear strength (ILSS).

Table 1. Mechanical properties testing of C/C and SiCNW-C/C composites.

Samples	Out-Of-Plane Compression		In-Plane Compression		ILSS
	σ (MPa)	E (GPa)	σ (MPa)	E (GPa)	τ (MPa)
C/C	68.18	2.72	90.84	1.99	11.36
SiCNW-C/C	223.13	4.05	190.06	4.21	32.94
Increment (%)	249.3%	48.9%	109.2%	111.6%	190.0%

3.6. Fracture Surfaces

In order to understand the reinforcement mechanism of C/C composites in in-situ grown SiCNW, the fracture surfaces of C/C and SiCNW-C/C composites were observed by SEM. Figure 7 shows the morphology of the fracture surface in C/C and SiCNW-C/C composites after compressive stress and ILSS. When the compressive stress was loaded on composite samples, the compressive strength mainly depended on the matrix cohesion [37]. In Figure 7a, the fracture surface of C/C was characterized by a rough structure accompanying bundles of fiber pull-out. As for SiCNW-C/C—shown in Figure 7d—the fracture surface was flat and nearly absent. This fracture surface could be attributed to the strongly enhanced cohesion in the matrix and also the powerful mechanical interlocking at the F/M and M/M interfaces, which allowed destructive cracks to extend into and through CFs (carbon fibers), without interfacial debonding (this has been detailed in Figure 8). In Figure 7b, the fracture surface of the C/C composites showed lots of stepwise fractured pyrolytic carbon (PyC) and many holes. The zoomed-in area in the top right corner of Figure 7b shows a step-shaped fracture along the fiber axis. These fracture steps resulted from the annular cracks that supplied paths for the spreading and linking up of destructive cracks, leading to the formation of stepwise PyC. There were no obvious stepwise fractures or holes but many SiC nanowires in the SiCNW-C/C, as shown in Figure 7e, indicating that SiCNWs could efficiently impede the propagation of destructive cracks along defects of the PyC matrix. When the interlaminar shear stress was loaded on composite samples, the shear strength mainly depended on both the matrix cohesion and F/M interfacial bonding strength. In Figure 7c, the smooth and clean fiber surface suggested a serious matrix delaminating between carbon fibers and the PyC matrix in the C/C composite, implying that primary fracture behaviors could be attributed to a typical delamination failure, which was similar to the failure mode observed in the compression test. As for the SiCNW-C/C—shown in Figure 7f—matrix delaminating was inhibited and many damaged carbon fibers could be observed in the shearing fracture surface, indicating that the interfacial strength between fibers and matrix was enhanced. In addition, faulted SiC nanowire can also be seen from the inset of Figure 7f.

SEM investigations—shown in Figure 8—into fracture surfaces of two kinds of composites, perpendicular to fiber direction, were provided to further explore SiC reinforcing mechanisms. As shown in Figure 8a, some annular cracks and small crazing (enlarged SEM image inset) could be detected easily, which implied a poor F/M and M/M interface bonding in C/C composites.

Compared with C/C, the presence of SiC—shown in Figure 8b—in SiCNW-C/C composites could effectively repair this crazing and reduce annular cracks. Corresponding EDS spectrums—shown in Figure 8c,d—could also prove this. According to the analysis above, failure mechanism schematic modeling of two kinds of composites, during the loading process, was established, as shown by the inset pictures in the left corner of Figure 8a,b.

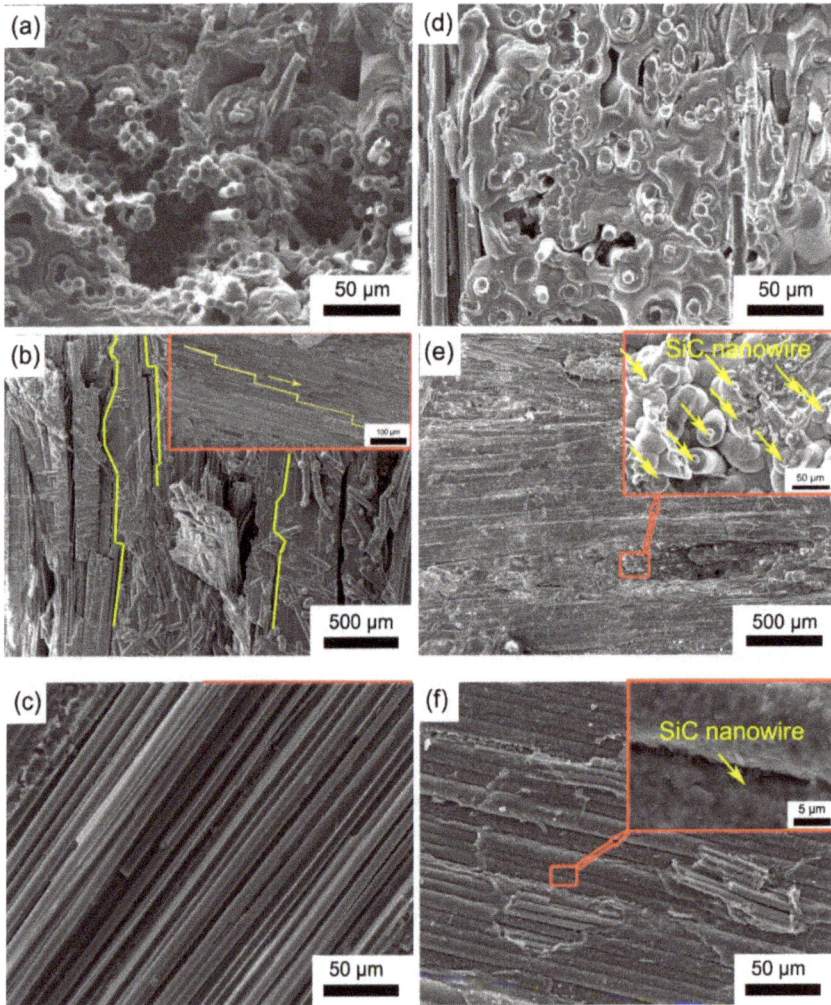

Figure 7. SEM images of compressive fractures and ILSS surfaces for the C/C (**a–c**) and SiCNW-C/C (**d–f**) composites.

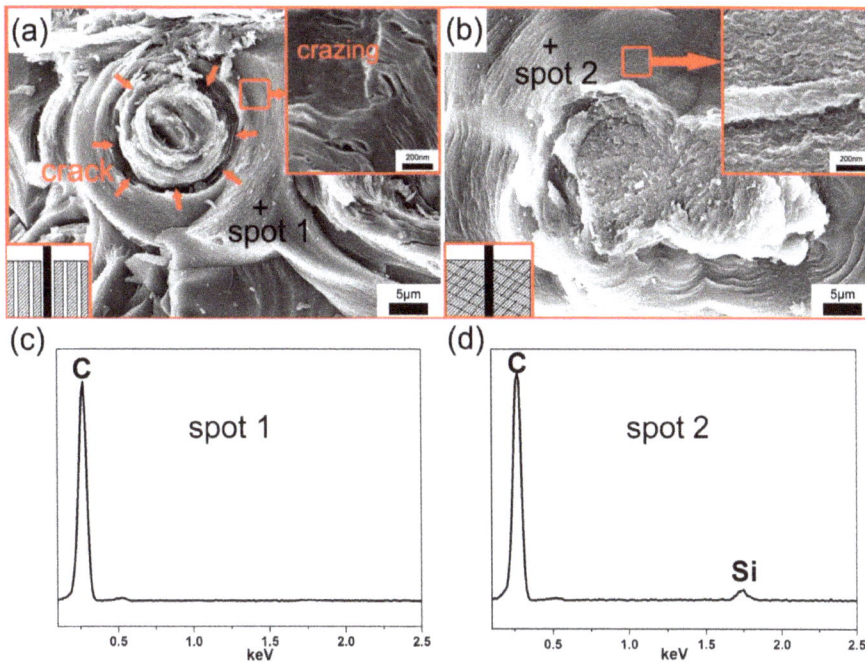

Figure 8. SEM images and EDS spectrums of the representative failure specimens from C/C (**a**,**c**) and SiCNW-C/C (**b**,**d**) composites, perpendicular to fiber direction.

4. Conclusions

SiCNW-C/C composites were obtained by a combination of sol-gel and chemical vapor infiltration processes. SiCNWs were in-situ synthesized inside of the C/C composites. No catalyst was introduced in the whole preparation process, which was environmentally friendly and cost-saving. The initial oxidation decomposition temperatures of C/C, modified by SiCNWs, was reduced by about 150 °C. Mechanical property tests indicated that there was a significant increase in the out-of-plane and in-plane compressive strengths, as well as, interlaminar shear strength of the C/C composites. SiCNWs improved simultaneously the F/M and M/M interfaces. SiCNWs not only efficiently decreased defects between interfaces, but also prevented crack growth. Hence, this work might open up a possibility to produce SiC-reinforced C/C composites with excellent mechanical strength, ductility, and toughness to replace traditional C/C in industries.

Author Contributions: Conceptualization, H.L. (Hongjiao Lin) and H.L. (Hejun Li); Methodology, H.L. (Hejun Li); Validation, H.L. (Hongjiao Lin), Q.S. and X.S.; Formal Analysis, H.L. (Hongjiao Lin); Investigation, T.F.; Data Curation, H.L. (Hongjiao Lin); Writing-Original Draft Preparation, H.L. (Hongjiao Lin); Writing-Review & Editing, Q.S. and T.F.; Supervision, X.S.; Project Administration, L.G.; Funding Acquisition, H.L. (Hejun Li).

Acknowledgments: This work was supported by the National Natural Science Foundation of China (grant nos. 51772247 and 5172780072), the Creative Research Foundation of Science and Technology on Thermostructural Composite Materials Laboratory (grant no. 6142911050217), and the National Science Basic Research Plan in the Shaanxi Province of China (grant no. 2017JM5098).

Conflicts of Interest: The authors declare no conflict of interest.

References

1. Xie, J.B.; Liang, J.; Fang, G.D.; Chen, Z. Effect of needling parameters on the effective properties of 3D needled C/C-SiC composites. *Compos. Sci. Technol.* **2015**, *117*, 69–77. [CrossRef]

2. Poitrimolt, M.; Cheikh, M.; Bernhart, G.; Velay, V. Characterisation of the transverse mechanical properties of carbon/carbon composites by spherical indentation. *Carbon* **2014**, *66*, 234–245. [CrossRef]

3. Huo, C.X.; Guo, L.J.; Wang, C.C.; Kou, G.; Song, H.T. Microstructure and ablation mechanism of SiC-ZrC-Al₂O₃ coating for SiC coated C/C composites under oxyacetylene torch test. *J. Alloys Compd.* **2018**, *735*, 914–927. [CrossRef]

4. Fan, X.M.; Yin, X.W.; Cao, X.Y.; Chen, L.Q.; Cheng, L.F.; Zhang, L.T. Improvement of the mechanical and thermophysical properties of C/SiC composites fabricated by liquid silicon infiltration. *Compos. Sci. Technol.* **2015**, *115*, 21–27. [CrossRef]

5. Chen, J.; Tian, C.; Luan, X.G.; Xiao, P.; Xiong, X. The mechanical property and resistance ability to atomic oxygen corrosion of boron modified carbon/carbon composites. *Mater. Sci. Eng. A* **2014**, *610*, 126–131. [CrossRef]

6. Lu, X.F.; Xiao, P.; Chen, J.; Long, Y. Oxidation behavior of C/C composites with the fibre/matrix interface modified by carbon nanotubes grown in situ at low temperature. *Corros. Sci.* **2012**, *55*, 20–25. [CrossRef]

7. Park, S.J.; Seo, M.K.; Lee, D.R. Studies on the mechanical and mechanical interfacial properties of carbon-carbon composites impregnated with an oxidation inhibitor. *Carbon* **2003**, *41*, 2991–3002. [CrossRef]

8. Chen, J.; Xiao, P.; Xiong, X. The mechanical properties and thermal conductivity of carbon/carbon composites with the fiber/matrix interface modified by silicon carbide nanofibers. *Mater. Des.* **2015**, *84*, 285–290. [CrossRef]

9. Li, R.; Lachman, N.; Florin, P.; Wagner, H.D.; Wardle, B.L. Hierarchical carbon nanotube carbon fiber unidirectional composites with preserved tensile and interfacial properties. *Compos. Sci. Technol.* **2015**, *117*, 139–145. [CrossRef]

10. Wu, S.; Liu, Y.Q.; Ge, Y.C.; Ran, L.P.; Peng, K.; Yi, M.Z. Surface structures of PAN-based carbon fibers and their influences on the interface formation and mechanical properties of carbon-carbon composites. *Compos. Part A* **2016**, *90*, 480–488. [CrossRef]

11. Song, Q.; Li, K.Z.; Li, H.L.; Li, H.J.; Ren, C. Grafting straight carbon nanotubes radially onto carbon fibers and their effect on the mechanical properties of carbon/carbon composites. *Carbon* **2012**, *50*, 3949–3952. [CrossRef]

12. Lu, X.F.; Xiao, P. Preparation of in situ grown silicon carbide nanofibers radially onto carbon fibers and their effects on the microstructure and flexural properties of carbon/carbon composites. *Carbon* **2013**, *59*, 176–183. [CrossRef]

13. Dai, W.; Yu, J.H.; Liu, Z.D.; Wang, Y.; Song, Y.Z.; Lyu, J.L.; Bai, H.; Nishimura, K.; Jiang, N. Enhanced thermal conductivity and retained electrical insulation for polyimide composites with SiC nanowires grown on graphene hybrid fillers. *Compos. Part A* **2015**, *76*, 73–81. [CrossRef]

14. Dong, R.H.; Yang, W.S.; Wu, P.; Hussain, M.; Xiu, Z.Y.; Wu, G.H.; Wang, P.P. Microstructure characterization of SiC nanowires as reinforcements in composites. *Mater. Charact.* **2015**, *103*, 37–41. [CrossRef]

15. Zhang, Y.L.; Hu, M.; Qin, X.G.; Song, X.G. The influence of additive content on microstructure and mechanical properties on the C-sf/SiC composites after annealed treatment. *Appl. Surf. Sci.* **2013**, *279*, 71–75.

16. Yang, W.; Araki, H.; Tang, C.C.; Thaveethavorn, S.; Kohyama, A.; Suzuki, H.; Noda, T. Single-crystal SiC nanowires with a thin carbon coating for stronger and tougher ceramic composites. *Adv. Mater.* **2005**, *17*, 1519–1523. [CrossRef]

17. Wu, R.B.; Zhou, K.; Yue, C.Y.; Wei, J.; Pan, Y. Recent progress in synthesis, properties and potential applications of SiC nanomaterials. *Prog. Mater. Sci.* **2015**, *72*, 1–60. [CrossRef]

18. Li, J.; Zhu, X.L.; Ding, P.; Chen, Y.P. The synthesis of twinned silicon carbide nanowires by a catalyst-free pyrolytic deposition technique. *Nanotechnology* **2009**, *20*, 145602. [CrossRef] [PubMed]

19. Dong, Z.J.; Meng, J.; Zhu, H.; Yuan, G.M.; Cong, Y.; Zhang, J.; Li, X.K.; Westwood, A. Synthesis of SiC nanowires via catalyst-free pyrolysis of silicon-containing carbon materials derived from a hybrid precursor. *Ceram. Int.* **2017**, *43*, 11006–11014. [CrossRef]

20. Lin, H.J.; Li, H.J.; Qu, H.Y.; Li, L.; Shi, X.H.; Guo, L.J. In situ synthesis of SiOC ceramic nanorod-modified carbon/carbon composites with sol-gel impregnation and CVI. *J. Sol-Gel Sci. Technol.* **2015**, *76*, 11–18. [CrossRef]

21. Luna, L.E.; Ophus, C.; Johanson, J.; Maboudian, R.; Carraro, C. Demonstration of hexagonal phase silicon carbide nanowire arrays with vertical alignment. *Cryst. Growth Des.* **2016**, *16*, 2887–2892. [CrossRef]

22. Han, M.K.; Yin, X.W.; Duan, W.Y.; Ren, S.; Zhang, L.T.; Cheng, L.F. Hierarchical graphene/SiC nanowire networks in polymer-derived ceramics with enhanced electromagnetic wave absorbing capability. *J. Eur. Ceram. Soc.* **2016**, *36*, 2695–2703. [CrossRef]

23. Zhang, X.D.; Huang, X.X.; Wen, G.W.; Geng, X.; Zhu, J.D.; Zhang, T.; Bai, H.W. Novel SiOC nanocomposites for high-yield preparation of ultra-large-scale SiC nanowires. *Nanotechnology* **2010**, *21*, 385601. [CrossRef] [PubMed]

24. Schoell, S.J.; Sachsenhauser, M.; Oliveros, A.; Howgate, J.; Stutzmann, M.; Brandt, M.S.; Frewin, C.L.; Saddow, S.E.; Sharp, I.D. Organic functionalization of 3C-SiC surfaces. *ACS Appl. Mater. Interfaces* **2013**, *5*, 1393–1399. [CrossRef] [PubMed]

25. Pan, J.M.; Yan, X.H.; Cheng, X.N.; Shen, W.; Li, S.X.; Cai, X.L. In situ synthesis and electrical properties of porous SiOC ceramics decorated with SiC nanowires. *Ceram. Int.* **2016**, *42*, 12345–12351. [CrossRef]

26. Zeng, Y.; Xiong, X.; Wang, D.N.; Wu, L. Residual thermal stresses in carbon/carbon-Zr-Ti-C composites and their effects on the fracture behavior of composites with different preforms. *Carbon* **2015**, *81*, 597–606. [CrossRef]

27. Meng, A.; Li, Z.J.; Zhang, J.L.; Gao, L.; Li, H.J. Synthesis and raman scattering of beta-SiC/SiO$_2$ core-shell nanowires. *J. Cryst. Growth* **2007**, *308*, 263–268. [CrossRef]

28. Chen, Z.M.; Ma, J.P.; Yu, M.B.; Wang, J.N.; Ge, W.K.; Woo, P.W. Light induced luminescence centers in porous SiC prepared from nano-crystalline SiC grown on Si by hot filament chemical vapor deposition. *Mater. Sci. Eng. B-Solid* **2000**, *75*, 180–183. [CrossRef]

29. Wu, R.B.; Yang, G.Y.; Gao, M.X.; Li, B.S.; Chen, J.J.; Zhai, R.; Pan, Y. Growth of SiC Nanowires from NiSi Solution. *Cryst. Growth Des.* **2009**, *9*, 100–104. [CrossRef]

30. Li, Z.J.; Zhang, J.L.; Meng, A.; Guo, J.Z. Large-area highly-oriented SiC nanowire arrays: Synthesis, Raman, and photoluminescence properties. *J. Phys. Chem. B* **2006**, *110*, 22382–22386. [CrossRef] [PubMed]

31. Li, J.Y.; Zhang, Y.F.; Zhong, X.H.; Yang, K.Y.; Meng, J.; Cao, X.Q. Single-crystalline nanowires of SiC synthesized by carbothermal reduction of electrospun PVP/TEOS composite fibres. *Nanotechnology* **2007**, *18*, 3999–4002. [CrossRef]

32. Mazo, M.A.; Nistal, A.; Caballero, A.C.; Rubio, F.; Rubio, J.; Oteo, J.L. Influence of processing conditions in TEOS/PDMS derived silicon oxycarbide materials. Part 1: Microstructure and properties. *J. Eur. Ceram. Soc.* **2013**, *33*, 1195–1205. [CrossRef]

33. Liang, T.A.; Li, Y.L.; Su, D.; Du, H.B. Silicon oxycarbide ceramics with reduced carbon by pyrolysis of polysiloxanes in water vapor. *J. Eur. Ceram. Soc.* **2010**, *30*, 2677–2682. [CrossRef]

34. Chen, J.H.; Liu, W.N.; Yang, T.; Li, B.; Su, J.D.; Hou, X.M.; Chou, K.C. A facile synthesis of a three-dimensional flexible 3C-SiC sponge and its wettability. *Cryst. Growth Des.* **2014**, *14*, 4624–4630. [CrossRef]

35. Lin, L.W. Synthesis and optical property of large-scale centimetres-long silicon carbide nanowires by catalyst-free CVD route under superatmospheric pressure conditions. *Nanoscale* **2011**, *3*, 1582–1591. [CrossRef] [PubMed]

36. Wu, R.B.; Li, B.S.; Gao, M.X.; Chen, J.J.; Zhu, Q.M.; Pan, Y. Tuning the morphologies of SiC nanowires via the control of growth temperature, and their photoluminescence properties. *Nanotechnology* **2008**, *19*, 335602. [CrossRef] [PubMed]

37. Feng, L.; Li, K.Z.; Xue, B.; Fu, Q.G.; Zhang, L.L. Optimizing matrix and fiber/matrix interface to achieve combination of strength, ductility and toughness in carbon nanotube-reinforced carbon/carbon composites. *Mater. Des.* **2017**, *113*, 9–16. [CrossRef]

nanomaterials

MDPI

Article

Long-Term Mechanical Behavior of Nano Silica Sol Grouting

Dongjiang Pan [1], Nong Zhang [1,*], Chenghao Zhang [1], Deyu Qian [1,*], Changliang Han [1] and Sen Yang [1,2]

1   Key Laboratory of Deep Coal Resource Mining, School of Mines, China University of Mining and Technology, Xuzhou 221116, China; cumtpdj@163.com (D.P.); zhangchenghao421@126.com (C.Z.); hclayor@126.com (C.H.); yangsen91811@163.com (S.Y.)
2   Department of Energy and Mineral Engineering, G3 Center and Energy Institute, Pennsylvania State University, PA 16802, USA
*   Correspondence: zhangnong@cumt.edu.cn (N.Z.); tbh078@cumt.edu.cn (D.Q.); Tel.: +86-516-8359-0502 (N.Z. & D.Q.)

Received: 29 November 2017; Accepted: 9 January 2018; Published: 16 January 2018

Abstract: The longevity of grouting has a significant effect on the safe and sustainable operation of many engineering projects. A 500-day experiment was carried out to study the long-term mechanical behavior of nano silica sol grouting. The nano silica sol was activated with different proportions of a NaCl catalyst and cured under fluctuating temperature and humidity conditions. The mechanical parameters of the grout samples were tested using an electrohydraulic uniaxial compression tester and an improved Vicat instrument. Scanning electron microscope, X-ray diffraction, and ultrasonic velocity tests were carried out to analyze the strength change micro-mechanism. Tests showed that as the catalyst dosage in the grout mix is decreased, the curves on the graphs showing changes in the weight and geometric parameters of the samples over time could be divided into three stages, a shrinkage stage, a stable stage, and a second shrinkage stage. The catalyst improved the stability of the samples and reduced moisture loss. Temperature rise was also a driving force for moisture loss. Uniaxial compressive stress-strain curves for all of the samples were elastoplastic. The curves for uniaxial compression strength and secant modulus plotted against time could be divided into three stages. Sample brittleness increased with time and the brittleness index increased with higher catalyst dosages in the latter part of the curing time. Plastic strength-time curves exhibit allometric scaling. Curing conditions mainly affect the compactness, and then affect the strength.

Keywords: nano silica sol; long-term mechanical tests; fluctuating temperature-humidity conditions; micro-mechanism

1. Introduction

Nano silica sol is a kind of grouting material made up of nanometer particles that is highly transmissive [1]; it has been used in the geotechnical field in recent years [2]. Jurinak [3] developed a practical fluid-flow control system based on colloidal silica gel for oilfield reservoirs. Persoff et al. [4] studied permanent barrier systems formed by colloidal silica and polysiloxane for contaminant isolation. Saito et al. [5] proposed the use of silica sol to maintain overburden stability during shield tunneling. Butron et al. [6,7] tested the mechanical properties and failure of silica sol used for rock grouting for periods of up to half a year. Funehag et al. [8–10] designed a grouting procedure for a tunnel using silica sol. McCartney et al. [11] evaluated the formation of hydraulic barriers for secondary containment through the permeation of colloidal silica grout. Gallagher et al. [12] studied a passive site stabilization technique for nondisruptive mitigation of liquefaction risk using colloidal silica. Wang et al. [13] used silica sol to seal phreatic pore water and fractured confined groundwater in the Xiao Jihan Coal Mine, China.

Because nano silica sol is hydrolyzed mineral slurry, both eco-friendly and non-polluting, it has a promising future for use in underground reservoir storage, CO_2 storage, nuclear waste storage, and for use as a sealant in dams, mines, and other engineered structures.

Although the service lives of geotechnical engineering structures can be very long, abundant data on the long-term mechanical properties of nano silica sol are still lacking. However, the longevity of grouting has a significant effect on the safe and sustainable operation of many engineering projects. For silica sol, the influence of the environment in which the sol cures, and the proportion of catalyst used on the long-term mechanical properties of the sol are important. Butrón et al. [6] studied the strength of different batches of silica sol after curing it under different temperatures and relative humidities for 6 months. However, these experiments were not long enough for many engineering purposes because the durability of grouting projects is commonly evaluated for three time periods. These are short-term, one–two months, medium term, two months to one year, and long term, and one year to three years or longer [14].

For this paper, the environmental conditions used to test nano silica sol curing simulated those present in an underground coal mine. The periodic variations of temperature and humidity in the atmosphere on the surface directly affect the microclimate in a mine [15,16]. Influenced by the temperature fluctuations of the mine airflow, the rock surrounding an underground roadway also experiences periodic temperature variations. This temperature field can have a radius of up to 30 or 40 m. Owing to high humidity in the return air, evaporation of water, and other causes, a relatively large number of roadways can experience high humidity for months at a time. To study how the mechanical properties of nano silica sol mixed with different proportions of catalyst change over the long term in an environment of fluctuating temperature and humidity, a 500-day test of the nano silica sol's mechanical properties was performed. Simultaneously, Scanning electron microscope (SEM), X-ray diffraction (XRD), and ultrasonic velocity tests were carried out to analyze the strength change micro-mechanism of samples. The objectives of this work are to evaluate the long-term stability of silica gel and the further measures can be taken to prolong the stability based on the study.

2. Materials and Methods

2.1. Materials

Silica sol is a stable liquid containing individual silica particles 8–12 nm in diameter [17]. The silica sol used for this study was provided by BASF HOCK Mining Chemical (China) Company Limited (Jining, China). Silica sol has an electric double layer structure, as shown in Figure 1. When ionic sodium is added, the electric double layer becomes thinner and the silica particles connect to form a gel. Some of the physical and chemical parameters [2,17] of the silica sol and NaCl catalyst are listed in Table 1. Gel time of silica sol to catalyst of 4:1, 7:1, and 9:1 (by volume) is 10 min, 2.67 h, and 9.33 h, respectively.

Table 1. Basic parameters of the hydrolyzed silica sol and catalyst used in this study.

Properties	Silica Sol	Catalyst
Viscosity	~10 mPa·s	~1 mPa·s
Density	1.1 kg/L	1.07 kg/L
pH	10	7
Concentration (% by weight)	SiO_2 15%	NaCl 10%

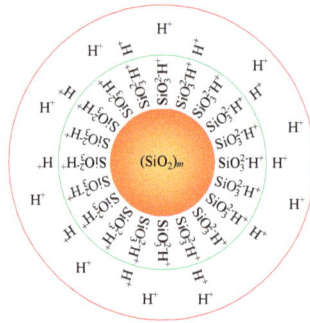

Figure 1. Electric double layer structure of nano silica sol.

2.2. Sample Preparation and Curing Conditions

The samples to be cured are prepared with volumetric proportions of silica sol to catalyst of 4:1, 7:1, and 9:1. The silica sol and catalyst are blended at a mixing speed of 100 revolutions per minute for 90 s utilizing RW 20 digital overhead stirrer (IKA® Works Guangzhou, Guangzhou, China). The mixture is poured into a mold, a 100 mm long transparent acrylic tube with an inner diameter of 50 mm. The next day, the samples are unmolded and are put into purpose-made curing boxes.

The curing boxes are designed to simulate the temperature and humidity of the rock surrounding a roadway, mimicking the high humidity and periodic temperature fluctuations in an underground mine. Water is poured into the bottom of each curing box and is replenished regularly to ensure a water depth of 5 mm. A foam board is immersed in the water and the samples are placed on the foam board. The temperature of the curing boxes is in natural state of a lab. An RC-4HC temperature-humidity recorder (Hangzhou SinoMeasure Automation technology Co., Ltd., Hangzhou, China) is used to monitor and record the temperature and relative humidity in the curing box. A schematic diagram of the samples, the temperature-humidity recorder, and a curing box is shown in Figure 2.

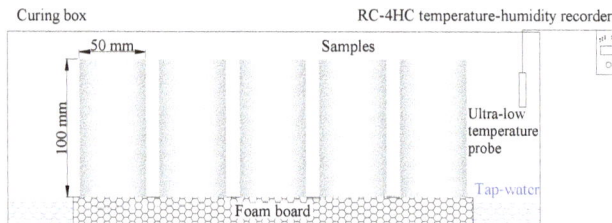

Figure 2. Schematic diagram of the samples, temperature-humidity recorder, and curing box used for this study. Water is added every month to maintain the depth of the water layer at 5 mm.

The samples were prepared on 29 April 2016. Temperature and relative humidity data from the preparation date to 11 September 2017 (500 days) are shown in Figure 3.

The temperature and humidity fluctuated over both long and short periods. For the short period fluctuations, the temperature changes are opposite those in the relative humidity. Overall, the relative humidity was maintained at 50–95% and humidity was relatively high, 70–90%, most of the time. The experiment was performed in Xuzhou City, lat 34.26° N long 117.20° E. The average daily temperature there is lower between November and February and higher between June and September. For the time intervals 0–150 days and 400–500 days of the test, the average daily temperature is usually higher than 20 °C; for the interval 200–400 days, the temperature is usually lower than 20 °C.

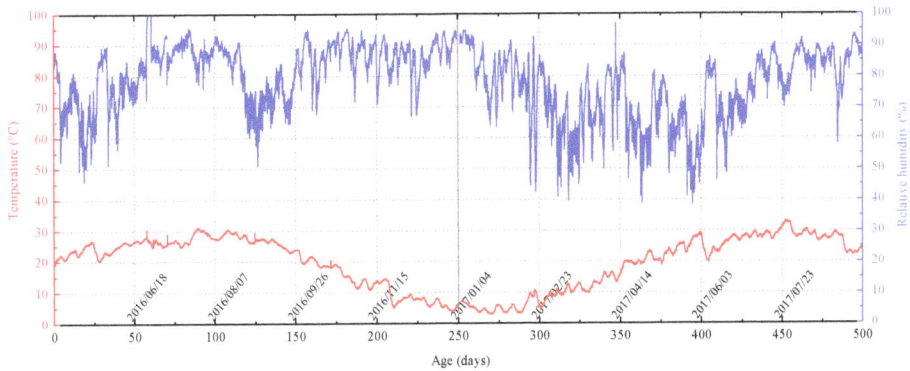

Figure 3. Temperature and relative humidity in the curing boxes between 29 April 2016 and 11 September 2017 (500 days). The temperature and humidity data show both long- and short-period fluctuations. The relative humidity was maintained at 50–95% and for much of the time was between 70% and 90%.

2.3. Testing Methods

The samples are weighed regularly on a precise electronic balance, the diameter and height measured with a vernier caliper, and the sample volumes calculated. The uniaxial compression strength (UCS), peak strain, secant modulus E_{50}, pre-peak absorption strain energy density (PASED), and other indexes are obtained regularly by uniaxial compression tests using a UTM5504 electrohydraulic servo test machine (Shenzhen Suns Technology Stock Co., Ltd., Shenzhen, China).

UCS and peak strain reflect the ability to resist damage; PASED is the internal force that causes sudden failure of a material. E_{50} is the ratio of the stress at one-half of the UCS to the corresponding strain. The ratio of UCS to peak strain is denoted as E_p. E_{50} and E_p both reflect the ability of a material to resist deformation. Nonlinear elasticity is expressed by the elastic modulus gradient, $(E_{50} - E_p)/E_{50}$. The ratio of strain energy density before and after peak stress defines the brittleness of the material. This ratio is called the brittleness index.

The ultimate shear stress (also known as plastic strength) of the grouting concretion was measured with an improved Vicat instrument (Beijing, China). Plastic strength, P_s, can be determined according to the equilibrium force relationships shown in Figure 3. The value of P_s can be calculated from Equation (1) [7,18]:

$$Ps = \frac{G \cos^2 \frac{\alpha}{2} \cot \frac{\alpha}{2}}{\pi \times h^2} \tag{1}$$

where P_s is the plastic strength of the grouting concretion during the solidification of the slurry, G is the total weight of the Vicat cone, α is the conical vertex angle, and h is the cone penetration depth.

3. Results

As shown in Figure 4, overall, the weight- and geometric parameter-time curves can be divided into three stages, a shrinkage stage, a stable stage, and a second shrinkage stage. Note that the three stages are more pronounced the lower the amount of catalyst used in the grout (or the higher the sol: catalyst ratio). In the first shrinkage stage, the weight, height, diameter, and volume of the sample decrease approximately linearly. The catalyst can apparently decrease the shrinkage rate and prolong the shrinkage stage. The graphs in Figure 4 also show that the catalyst enhances the stability of the samples and reduces moisture loss. In the stable stage, the weight and the geometric parameters remain almost unchanged. Because the stable stage takes place in the winter (Figure 2), the temperature is mainly around 8 °C and the lower temperature inhibits the sublimation of moisture. The smaller the amount of catalyst, the lower the stability. During the second shrinkage stage, the temperature has increased to 20–25 °C, enhancing moisture loss. During this stage, the weight and height of 4:1 samples decrease slightly, although the volume and diameter increases slightly. In general, the weight and geometric parameters for the 4:1 samples remain almost unchanged. The 7:1 and 9:1 samples, however, begin to shrink again although the shrinkage rate is slightly slower than during the first shrinkage stage.

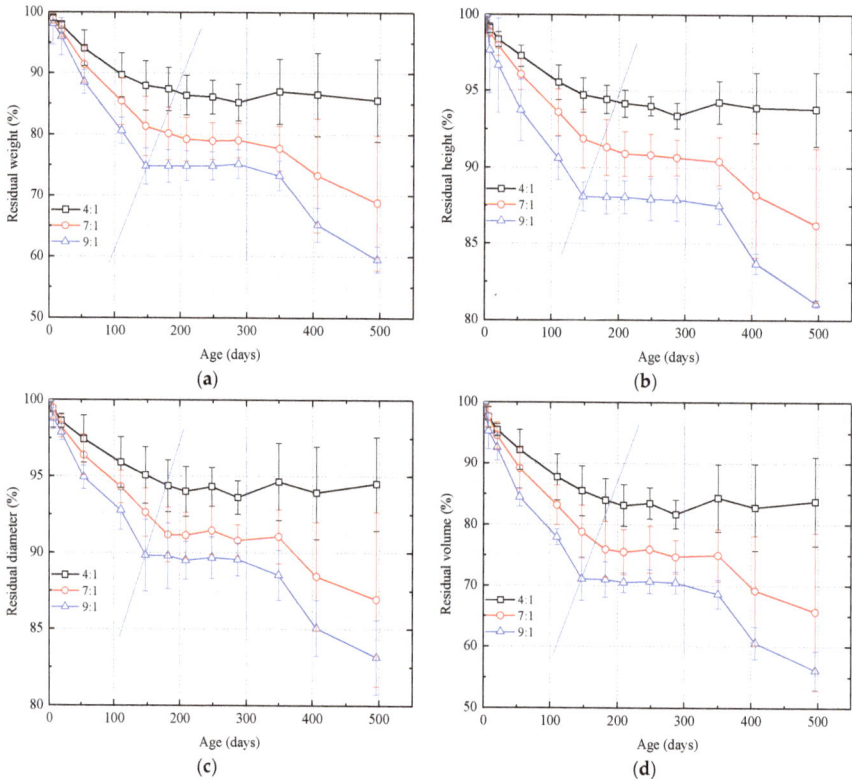

Figure 4. Graphs showing weight and geometric parameters vs. time for samples with different silica sol:catalyst ratios. (**a**) Residual weight; (**b**) Residual height; (**c**) Residual diameter; (**d**) Residual volume. The curves can be divided into three stages and the graphs show that the lower the amount of catalyst, the more pronounced the stages.

Figure 5 shows that the uniaxial compressive stress-strain curves for samples with all three silica sol:catalyst ratios are elastoplastic no matter the sample's age. The deformation can be divided into four phases. These are (1) a compaction phase at strains of 0% to 1%. The deduction is that the weight and volume decrease causing internal micropores to be generated and the micropores are then closed by the external forces. (2) An elastic phase during which the stress strain curve is approximately linear. This phase endures over a great range of strain percentages. (3) A plastic phase during which a crack is initiated in the sample and rapidly develops. The crack propagates and the sample is damaged by I-shaped tensile fracture. This stage is relatively short. (4) The post-peak stress phase, the sample is suddenly broken into pieces and flies apart almost instantaneously.

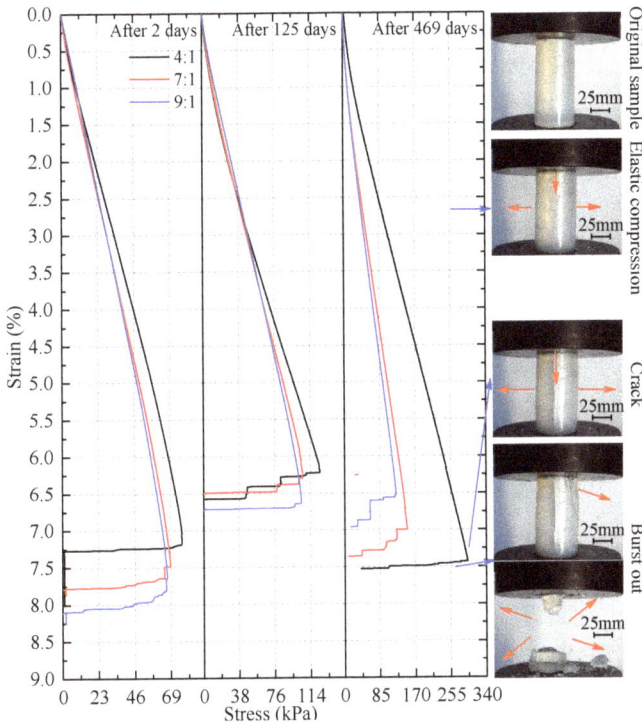

Figure 5. Uniaxial compressive stress-strain curves for samples with different proportions of silica sol and catalyst at three different times during the course of the experiment. The curves show that all the samples behaved as elastoplastic solids.

As shown in Figures 6 and 7, the way in which UCS and E_{50} change over time are similar, although the magnitudes of the changes are different. The curves can be divided into three stages, an ascending stage, a descending stage, and a second ascending stage. In the first ascending stage, the first 150 days, UCS and E_{50} increase. In the descending stage, from around day 150 to days 200–230, UCS and E_{50} gradually drop. In the second ascending stage, UCS and E_{50} have increased since the 200–230-day time and the rate of increase is higher than that during the first ascending stage. The curves also show that more catalyst causes UCS and E_{50} to reach higher values.

The samples' peak strain changed little with age and remained in the 4–8% range, only two percentage points above or below the average peak strain of 6%. This indicates that the samples could undergo significant deformation before they failed. The amount of catalyst also had little effect on peak strain.

Nonlinear elasticity of samples with different proportions of catalyst was between 0% and 18%, with an average of 8%. The results show that, before failure, the samples behave as linear elastic materials.

As shown in Figure 8, before day 400, the PASED curves are W-shaped and are basically unaffected by the amount of catalyst. The two troughs are around day 75 and day 230 and the peak is around day 125. After day 400, however, PASED increases rapidly and the effect of the amount of catalyst on PASED becomes significant. Higher silica sol:catalyst ratios cause PASED to increase to higher values, and the rate of increase is more rapid.

As illustrated by Figure 8b, before 400 days of curing, the brittleness index of the samples was essentially unchanged with an average value of 12.40, a high brittleness. The sol:catalyst ratio had almost no influence on the brittleness index. However after 400 days of curing, the greater the amount of catalyst used in the grout mixture, the higher the brittleness index.

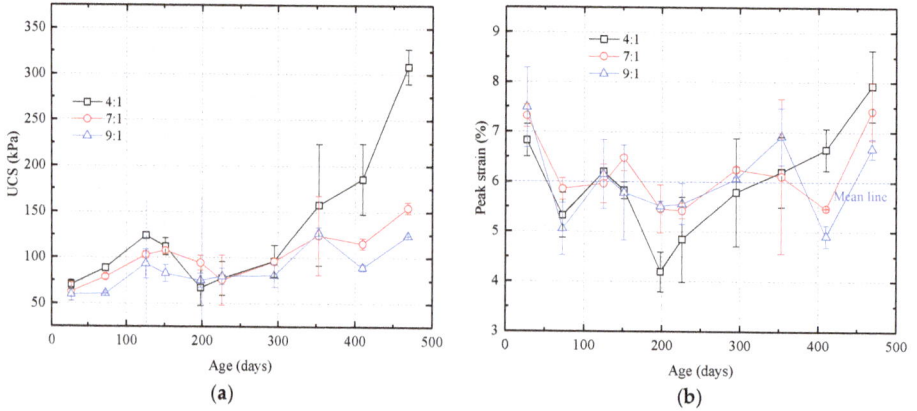

Figure 6. Uniaxial compression strength and peak strain vs. time curves for samples with different proportions of silica sol and catalyst. (**a**) UCS curves. The curves can be divided into an ascending stage, a descending stage, and a second ascending stage; (**b**) Peak strain curves. Note that the sample's age and the proportion of catalyst have little effect on the peak strain.

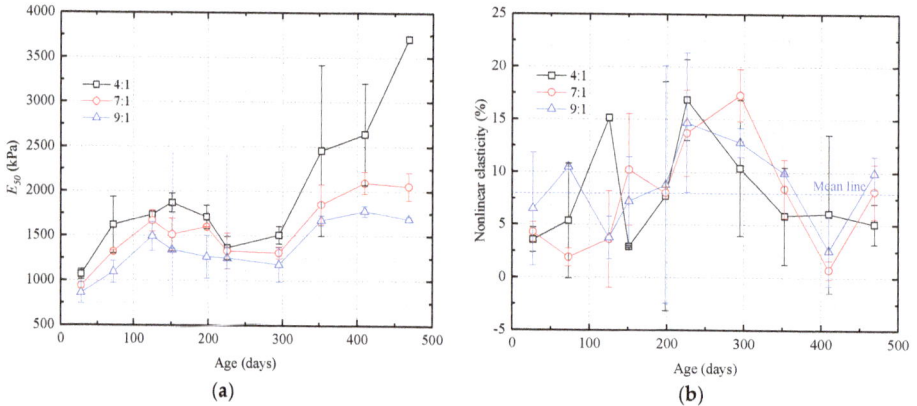

Figure 7. Secant modulus (E_{50}) and elastic modulus gradient vs. time curves for samples with different proportions of silica sol and catalyst. (**a**) E_{50} curves. The curves can be divided into an ascending stage, a descending stage, and a second ascending stage; (**b**) Nonlinear elasticity curves. Throughout the course of the experiment, the nonlinear elasticity remains low. Before failure, the samples behave as linear elastic materials.

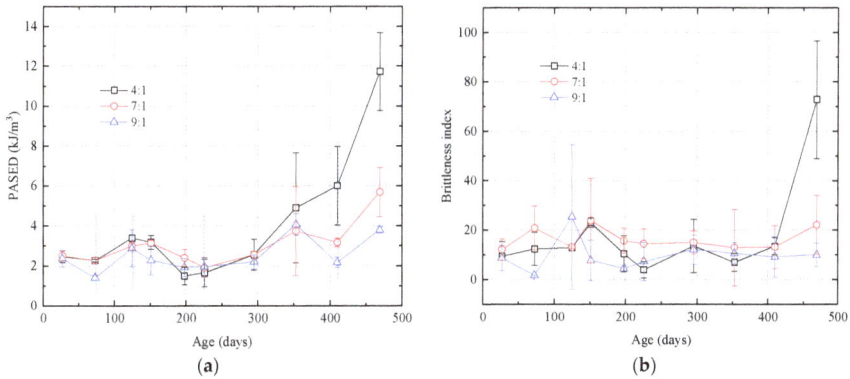

Figure 8. Graphs showing pre-peak absorption strain energy density (PASED) and brittleness index curves vs. time for samples with different silica sol:catalyst ratios. (**a**) PASED. The curves are W-shaped prior to 400 days. After 400 days, higher catalyst ratios result in significantly higher PASED; (**b**) Brittleness index. Samples with different silica sol:catalyst ratios have similar brittleness indexes before 400 days but after 400 days, samples with more catalyst have higher brittleness indexes.

As shown in Figure 9, the plastic strength-time curves for samples with different proportions of silica sol and catalyst exhibit allometric scaling. The equation for the line fit to the data in Figure 9 is shown below as Equation (2):

$$P_s = 22.444t^{0.452} \tag{2}$$

where, P_s is the plastic strength; t is the age.

Before 400 days, the effect of catalyst dosage on the grout's plastic strength is not clear. After 400 days, the less the amount of catalyst used, the greater the grout concretion's plastic strength.

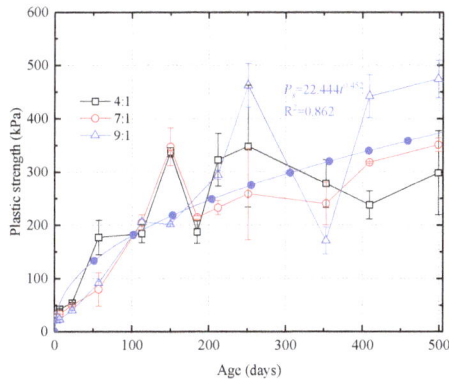

Figure 9. Plastic strength-time curves for samples with different proportions of silica sol and catalyst. The curves show allometric scaling for plastic strength.

4. Discussion

4.1. Weight- and Geometric Parameter-Time Law

The difference between this study and the study done by Axelsson (2006) is that in Axelsson's experiments, the curing temperature was 8 °C, and the samples were cured under three controlled relative humidities, 75, 95, and 100%. Axelsson's 180-day experiments showed that the greater the relative humidity, the less the drying shrinkage and that shrinkage tended to stabilize by day 180. In the experiment done for this paper, the relative humidity for the first 180 of the 500 days concentrated in the 70–90% range, and the temperature was usually between 20 and 30 °C. The drying shrinkage for our samples during this time is similar to that of Axelsson's samples and the shrinkage also tends to stabilize at about day 180. In addition, as shown in Figure 4, less catalyst in the grout mix results in higher shrinkage. These results show that 180 days may be the curing time needed for initial drying shrinkage stability under high relative humidity (greater than 70%) and either normal (20–30 °C) or low (8 °C) temperatures. In Xuzhou City, where this study was conducted, the 200–300-day study interval was in the winter (November to February, Figure 2) and the average high temperature was about 8 °C. These temperatures essentially extended Axelsson's (2006) experiment. In the 200–300-day period, the weight, height, diameter, and volume of the samples were fairly stable and this also indicates that 180 days might be the curing time for initial drying shrinkage stability. In this study's second shrinkage stage, temperature was back up to the 20–30 °C range, reactivating moisture loss. Temperature rise is a driving force for moisture loss.

4.2. Strength- and Secant Modulus-Time Law and Micro-Mechanism

Butrón et al. (2009) showed that the strength of silica sol grout increased over time in 180-day experiments. As illustrated in Figures 6 and 7, the values for UCS and secant modulus from samples in this study can be divided into an ascending stage, a descending stage, and a second ascending stage. The first ascending stage results are similar to Butrón et al.'s (2009) experimental results. The Butrón experiments also showed that samples that were cured for the same amount of time have greater strength when cured at higher temperatures under lower relative humidities. For the samples tested by this study, the weight and geometric parameters of the samples remained about the same in the descending stage, but in this study, the temperature decreased significantly and the relative humidity increased slightly. This suggests that the descending stage in this experiment was caused by the combined effects of temperature and humidity. In this study's second ascending stage, temperature increased, relative humidity decreased slightly, and UCS and the secant modulus increased. The brittleness index also increased.

At the same time of mechanical tests, microstructure, crystal characteristics and compactionness experiments were carried out to analyze the strength change micro-mechanism of samples. As shown in Figure 10, the microstructure properties of samples were inspected with a FEI Quanta™ 250 scanning electron microscope (SEM) (Hillsboro, OR, USA). There are few pores in the initial stage, and there is no pore on the surface of the intermediate and later samples. Only a small number of irregular particles adhere to the surface of the sample. As shown in Figure 11, D8 Advance X-ray diffraction (XRD) (Bruker Corporation, Karlsruhe, Germany) showed that the crystal structure of samples are basically unchanged. As shown in Figure 12, C61 Ultrasonic Non-metal Detector (Sinotesting Technology, Beijing, China) showed that ultrasonic velocity-time curves for samples are similar to UCS curves. The ultrasonic velocity represents the compactionness. These indicate that the curing conditions mainly affect the compactness, and then affect the strength.

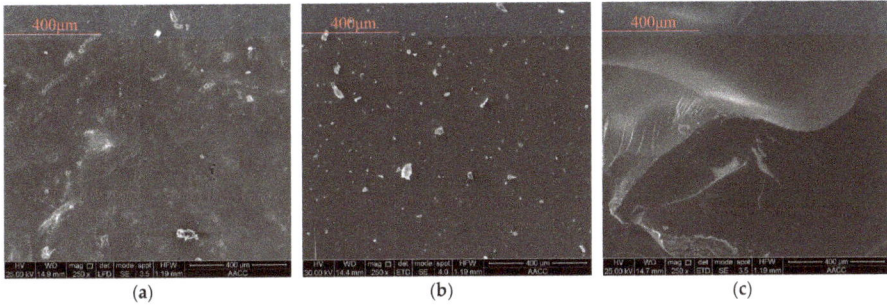

Figure 10. Scanning electron microscope (SEM) results for samples at three different times during the course of the experiment. (**a**) 2 d; (**b**) 206 d; (**c**) 500 d. The magnification of all is 250 times.

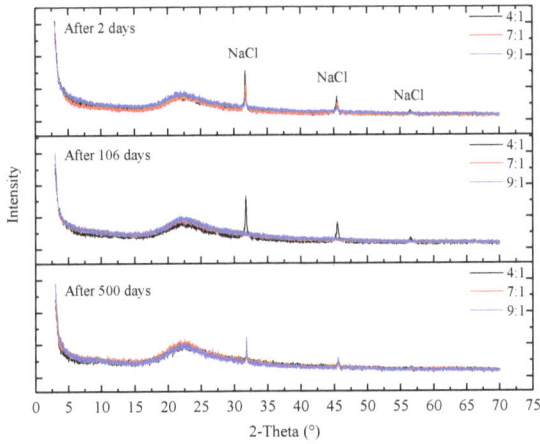

Figure 11. X-ray diffraction (XRD) results for samples with different proportions of silica sol and catalyst at three different times during the course of the experiment.

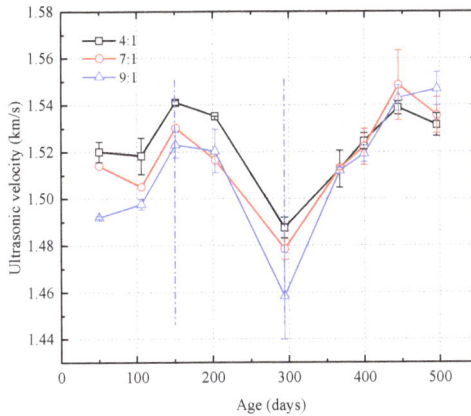

Figure 12. Ultrasonic velocity-time curves for samples with different proportions of silica sol and catalyst.

4.3. Plastic Strength-Time Law

When the Vicat cone is in static equilibrium, the calculated plastic strength includes shear strength from the grouting concretion on the sides of the cone and the grouting concretion's compressive strength. Therefore, the plastic strength should be slightly larger than the unconfined compressive strength. This study also proves this relationship. The study also shows that as curing time increases, the plastic strength scales allometrically.

5. Conclusions

To study the long-term mechanical behavior of nano silica sol grout prepared with different proportions of catalyst in an environment of fluctuating temperature and humidity, a 500-day experiment has been carried out. Simultaneously, SEM, XRD, and ultrasonic velocity tests were carried out to analyze the strength change micro-mechanism of samples. The main conclusions from this experiment are:

(1) The temperature and humidity fluctuate over both long and short periods. As the dosage of catalyst in the grout mix is decreased, the curves showing the changes in sample weight and sample width, height, and volume over time can be divided into three stages, a shrinkage stage, a stable stage, and a second shrinkage stage. Higher amounts of catalyst improve the stability of the samples and reduce moisture loss. Temperature rise is also a driving force for moisture loss.

(2) The uniaxial compressive stress-strain curves all show that the samples are elastoplastic. The deformation can be divided into four phases, a compaction phase, an elastic phase, a plastic phase, and a post-peak stress phase. The curves for the uniaxial compression strength and the secant modulus can be divided into an ascending stage, a descending stage, and a second ascending stage. Peak strain for the samples changed little with curing time. The PASED-time curves are W-shaped and are essentially unaffected by the amount of catalyst prior to 400 days. After 400 days, higher catalyst ratios increase the PASED values significantly. Sample brittleness increases with time and in the later stages of the experiment, the brittleness index increases with higher catalyst dosages.

(3) Plastic strength-time curves for samples with different proportions of catalyst exhibit allometric scaling. A consistent effect of catalyst dosage on plastic strength is not apparent prior to 400 days of curing but after 400 days, it is clear that when the grout mix contains less catalyst, the plastic strength of the grout is greater.

(4) The surfaces of samples are smooth and compact at different ages, substantially unchanging. The crystal structure of samples are basically unchanged. Ultrasonic velocity-time curves for samples are similar to UCS curves. These indicate that the curing conditions mainly affect the compactness, and then affect the strength.

Acknowledgments: This work was supported by National Key Research and Development Program of China (2017YFC0603001); National Natural Science Foundation of China (51674244, 51704277 and 51404251); the Priority Academic Program Development of Jiangsu Higher Education Institutions; Innovative Research Teams in University by Ministry of Education of China (IRT_14R55); Fundamental Research Funds for the Central Universities (2014XT01); and China Postdoctoral Science Foundation (2017M621874). We thank David Frishman, PhD, from Liwen Bianji, Edanz Group China (www.liwenbianji.cn/ac), for editing the English text of a draft of this manuscript.

Author Contributions: Dongjiang Pan and Nong Zhang conceived and designed the experiments; Chenghao Zhang performed the experiments; Dongjiang Pan and Deyu Qian analyzed the data; Sen Yang contributed analysis tools; Dongjiang Pan wrote the paper; Changliang Han and Deyu Qian modified the paper.

Conflicts of Interest: The authors declare no conflict of interest. The founding sponsors had no role in the design of the study; in the collection, analyses, or interpretation of data; in the writing of the manuscript, and in the decision to publish the results.

References

1. Morris, C.; Anderson, M.; Stroud, R.; Merzbacher, C.; Rolison, D. Silica sol as a nanoglue: Flexible synthesis of composite aerogels. *Science* **1999**, *284*, 622–624. [CrossRef] [PubMed]
2. Pan, D.; Zhang, N.; Han, C.; Yang, S.; Zhang, C.; Xie, Z. Experimental Study of Imbibition Characteristics of Silica Sol in Coal-Measure Mudstone Matrix. *Appl. Sci.* **2017**, *7*, 300. [CrossRef]
3. Jurinak, J.; Summers, L. Oilfield Applications of Colloidal Silica Gel. *SPE Prod. Eng.* **1991**, *6*, 406–412. [CrossRef]
4. Persoff, P.; Finsterle, S.; Moridis, G.; Apps, J.; Pruess, K.; Muller, S. *Injectable Barriers for Waste Isolation*; Technical Report; Lawrence Berkeley Lab: Berkeley, CA, USA, 1995; pp. 1–11. [CrossRef]
5. Zhang, F.; Zhu, H.; Fu, D. *Shield Tunnel*; China Communications Press: Beijing, China, 2004; pp. 260–261. ISBN 7-114-05174-3.
6. Butrón, C.; Axelsson, M.; Gustafson, G. Silica sol for rock grouting: Laboratory testing of strength, fracture behaviour and hydraulic conductivity. *Tunn. Undergr. Space Technol.* **2009**, *24*, 603–607. [CrossRef]
7. Axelsson, M. Mechanical tests on a new non-cementitious grout, silica sol: A laboratory study of the material characteristics. *Tunn. Undergr. Space Technol.* **2006**, *21*, 554–560. [CrossRef]
8. Funehag, J.; Gustafson, G. Design of grouting with silica sol in hard rock—New methods for calculation of penetration length, Part I. *Tunn. Undergr. Space Technol.* **2008**, *23*, 1–8. [CrossRef]
9. Funehag, J.; Gustafson, G. Design of grouting with silica sol in hard rock—New design criteria tested in the field, Part II. *Tunn. Undergr. Space Technol.* **2008**, *23*, 9–17. [CrossRef]
10. Butrón, C.; Gustafson, G.; Fransson, Å.; Funehag, J. Drip sealing of tunnels in hard rock: A new concept for the design and evaluation of permeation grouting. *Tunn. Undergr. Space Technol.* **2010**, *25*, 114–121. [CrossRef]
11. McCartney, J.; Nogueira, C.; Homes, D.; Zornberg, J. Formation of Secondary Containment Systems Using Permeation of Colloidal Silica. *J. Environ. Eng.* **2011**, *137*, 444–453. [CrossRef]
12. Hamderi, M.; Gallagher, P.; Lin, Y. Numerical Model for Colloidal Silica Injected Column Tests. *Vadose Zone J.* **2014**, *13*, 138–143. [CrossRef]
13. Wang, Q.; Li, H.; Zhu, N. Research on water plugging with nanoscale grouting material and high-property concrete in slope shaft. *Coal Min. Technol.* **2013**, *18*, 97–98. [CrossRef]
14. Cheng, X.; Zhang, F. *Grouting Construction and Effect Detection in Civil Engineering*; Tongji University Press: Shanghai, China, 1998; p. 78. ISBN 7-5608-1763-7.
15. Zhao, J.; Yao, Y. Research on periodically change of airflow temperature in coal mine. *China Coal* **2012**, *38*, 98–101. [CrossRef]
16. Qin, Y.; Song, H.; Wu, J.; Dong, Z. Numerical analysis of temperature field of surrounding rock under periodic boundary using Finite Volume Method. *J. China Coal Soc.* **2015**, *40*, 1541–1549. [CrossRef]
17. Pan, D.; Zhang, N.; Xie, Z.; Feng, X.; Kong, Y. Laboratory Testing of Silica Sol Grout in Coal Measure Mudstones. *Materials* **2016**, *9*, 940. [CrossRef] [PubMed]
18. Xiao, Z.; Liu, B.; Qiao, S.; Yang, X.; Wu, G. Experimental research on new grouting materials of acidic water glass-calcium carbonate. *Rock Soil Mech.* **2010**, *31*, 2829–2834. [CrossRef]

MDPI

St. Alban-Anlage 66

4052 Basel

Switzerland

Tel. +41 61 683 77 34

Fax +41 61 302 89 18

www.mdpi.com

Nanomaterials Editorial Office

E-mail: nanomaterials@mdpi.com

www.mdpi.com/journal/nanomaterials

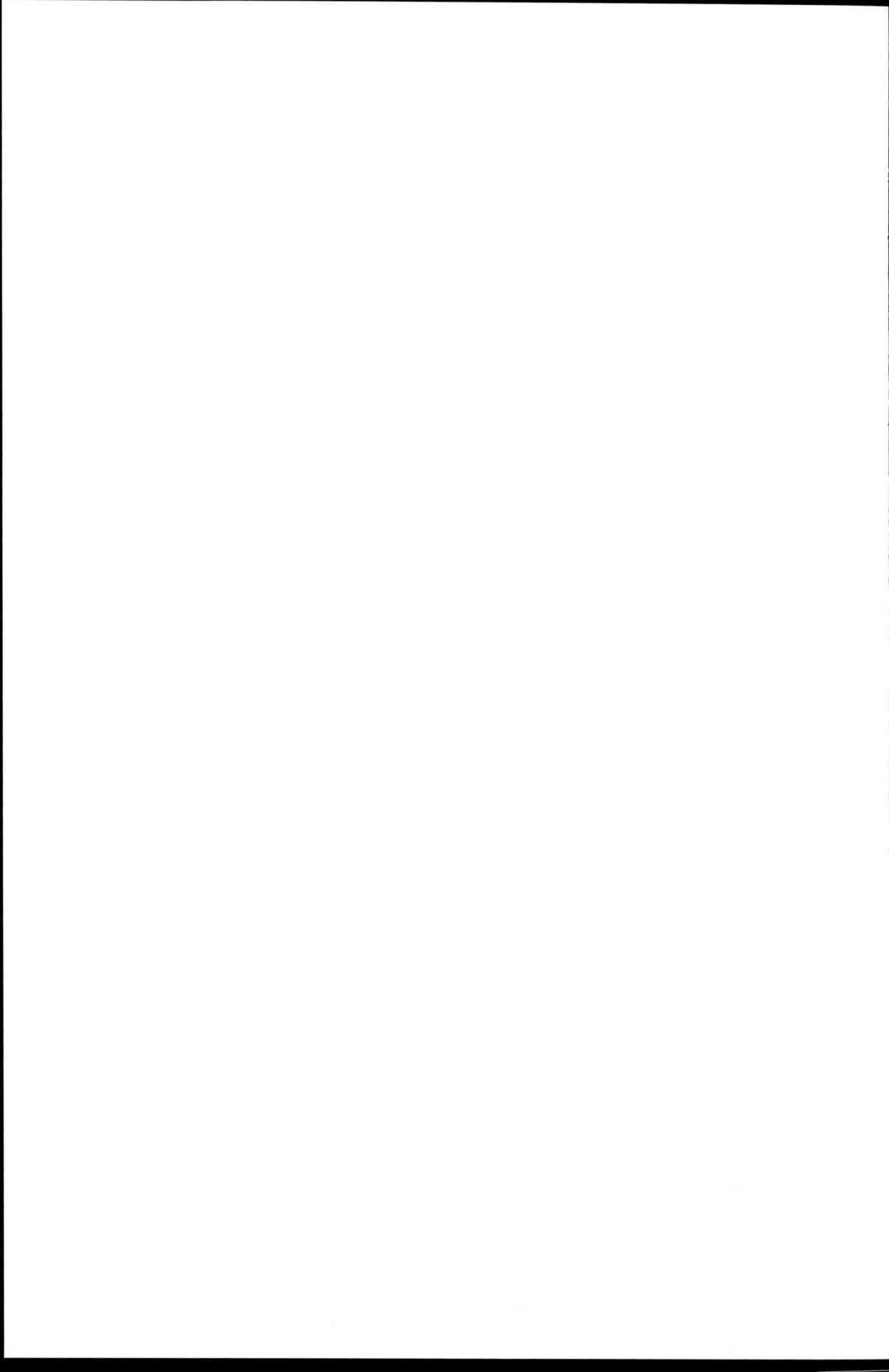

www.ingramcontent.com/pod-product-compliance
Lightning Source LLC
Chambersburg PA
CBHW051916210326
41597CB00033B/6165